INDUSTRIAL CHEMICAL CLEANING

James W. McCoy

Formerly Supervisor, Refinery Services
Chevron U.S.A., Inc.

Chemical Publishing Co.
New York, N.Y.

ISBN 0-8206-0305-8

Chemical Publishing Co., Inc.

Printed in the United States of America

Preface

This book has been written to provide information for chemists, process engineers, and management personnel who deal with industrial chemical cleaning contractors. Because of the specialized equipment required, chemical cleaning is seldom done in-house, but, even so, the client should be qualified to discuss the details of proposed procedures with the contractor when negotiating a contract. Furthermore, several of the methods discussed in the book are protected by patents, so they are best left to qualified contractors who have made the necessary licensing arrangements. As with any purchased service, however, it is to the purchaser's advantage to have a good understanding of the work to be done and the methods to be used. The principles and practice of industrial cleaning described here will enable any technically trained individual to understand chemical cleaning operations.

Chapter I contains a discussion of the sources and composition of various types of fouling deposits; the mechanism of corrosion; hazards that are encountered in chemical cleaning; and a brief description of the special equipment used by contractors. Chapters II, III, and IV contain detailed descriptions of the numerous chemicals used for industrial cleaning including alkalis, acids, corrosion inhibitors, and wetting agents, as well as some special agents. Several patented processes are also considered, as are nonaqueous solvents, foams, and a few nonchemical procedures, the most useful of which is high pressure water blasting.

Chapter V explores the principles and practice of the very important subject of passivation. Chapter VI deals with planning and contracting of chemical cleaning jobs with emphasis on safety and the proper disposal of hazardous wastes.

The next four chapters, VII-X, contain specific procedures for deoiling, cleaning, and passivating steam generators, heat exchangers, columns, reactors, and furnaces. The directions given here are intended as general information and should not be applied indiscriminately without the advice of a qualified professional contractor. Great care must be taken, for instance, to take account of specific metallurgical considerations to avoid major damage to equipment by incompatible chemicals.

Chapter XI includes laboratory procedures for measuring the strength of cleaning solutions, estimating solubilities, sampling fouling deposits, and evaluating industrial cleaners.

Although I must assume responsibility for any errors, I wish to thank my associate, Mr. Donald A. Alexander, for his thorough reading of the manuscript and his many suggestions that led to improvements in the work. He is widely recognized in the petroleum refining industry for his detailed knowledge of chemical cleaning processes and for his outstanding safety record as a supervisor of these inherently hazardous operations. Mr. Robert J. Hammond, who is unusually talented in piping layout work, kindly provided the piping diagrams. Finally, I am most grateful to my wife, Dolores, for her many contributions to the success of my books.

Richmond, California
May 31, 1983

James W. McCoy

Table of Contents

I

Principles of Chemical Cleaning

The formation of various types of scales, sludges, and deposits in process equipment of the petroleum refining and chemical processing industries occurs frequently. Some of these present serious problems—for example, the scaling of boiler tubes, or corrosive attack on the metals of which various types of reactors and similar vessels are constructed. Others are less significant—for instance, the phosphate sludges formed in boiler drums as the result of over-treatment with phosphate, or the formation of deposits of coke and salts on the floors of furnaces.

When such equipment is taken out of service, preferably on a scheduled shut-down for maintenance, but sometimes because of failure or loss of heat transfer capacity, the engineer in charge of operations often obtains a chemical analysis of any fouling material found. This is done for any of several reasons: to find whether the deposit resulted from corrosive attack on the vessel or from foreign material brought into the vessel; to indicate the reason for failure of equipment; or to assess the extent to which a vessel has been attacked during a certain period of operation. More often than not, however, all that is desired is an indication of the conditions necessary for the removal of the deposit from the site of formation. In this case, no analysis is required, but rather a report on the loosening effect of water, steam, detergents, dilute inhibited acids, alkalis, or other agents on the material.

Within the past 30 years it has become apparent that in-place cleaning of industrial equipment by means of chemical solutions is, in many instances, safer and more economical than the older, labor-intensive, mechanical methods. As the choice of cleaning agent is to some extent empirical, laboratory performance tests with representative samples are desirable to determine solubility in various solutions, the most effective temperature, and the efficiency to be expected in cleaning the fouled equipment. Specific directions for sampling are difficult to provide because of the many different circumstances encountered. Nevertheless, it is evident that the interpretation of analytical and solubility data depends upon a knowledge of the normal operation of process units and upon information concerning the location, consistency, distribution, and other features of the deposit. In vessels containing sections, baffles, or trays, for instance, the distribution of fouling is often significant, and samples from different locations may vary greatly in composition and solubility.

I.1 SOURCES AND COMPOSITION OF FOULING

Before any attempt is made to dissolve or loosen scales or deposits with a chemical solution, it is necessary to characterize the material as to its composition, which in some measure is related to its origin. For the purpose of a systematic discussion, fouling material is here arbitrarily classified by origin as water-formed, corrosion, microbiological, organic, or combustion deposits.

I.1.a Water-Formed Scales and Deposits

Scales and deposits are regularly encountered in aqueous cooling systems, mixers, solution tanks, boilers, and other steam generators, surface condensers, and in equipment in which heat is transferred from a process stream to water. The term scale describes a continuous, adherent layer of foreign material crystallized on the water side of a surface through which heat is exchanged. In principle, by adding certain chemicals the growth of scales can be inhibited and the insoluble particles can be dispersed in the recirculating water and removed by blowdown. Should the particles come out of suspension, however, they can accumulate as sludge in quiet sections of a boiler or cooling system. Deposit is a general term applied to more-or-less loose accumulations often found in less-turbulent sections of boilers, cooling systems, and water-treating facilities.

Scales are objectionable because of their insulating effect. In a boiler tube, for instance, they cause overheating and eventual failure of the metal. Deposits often cause plugging in critical areas such as waterwalls, waterwall headers, in blowdown lines, and in gauge glasses. Many different mineral structures have been identified in boiler scales by the methods of x-ray diffraction, electron diffraction, and polarizing microscopy. Examples of silicate scales are: acmite, $Na_2O \cdot Fe_2O_3 \cdot 4SiO_2$; analcite, $Na_2O \cdot Al_2O_3 \cdot 4SiO_2 \cdot 2H_2O$; pectolite, $Na_2O \cdot 4CaO \cdot 6SiO_2 \cdot H_2O$; serpentine, $3MgO \cdot SiO_2 \cdot 2H_2O$; sodalite, $Na_2O \cdot 3Al_2O_3 \cdot 6SiO_2 \cdot 2NaCl$; xonotlite, $5CaO \cdot 5SiO_2 \cdot H_2O$. When phosphate is used for internal treatment, ferric phosphate, $FePO_4$, basic magnesium phosphate, $Mg_3(PO_4)_2 \cdot Mg(OH)_2$, and hydroxyapatite, $Ca_{10}(PO_4)_6(OH)_2$, may also be precipitated, as well as the more common anhydrite, $CaSO_4$, and aragonite, $CaCO_3$. As already noted, the presence of these and other scales impedes the circulation of water and reduces heat transfer, both of which cause overheating and failure of tubes.

Scales and deposits form in a boiler because the compounds of which they are composed are insoluble under the conditions prevailing there. Two factors combine to make calcium salts especially troublesome: certain anhydrous calcium salts, notably the sulfate, decrease in solubility as temperature and pressure increase, and increasing temperature shifts the equilibrium of the following reaction to the right, causing $CaCO_3$ to precipitate.

$$Ca^{++} + 2HCO_3^{-} = CaCO_3 + CO_2 + H_2O \qquad \text{(I-1)}$$

In addition, hydrolysis of excess bicarbonate increases the concentration of hydroxyl ion, precipitating $Mg(OH)_2$, the solubility product of which is 5.5×10^{-12}. The solubility of $CaSO_4$ decreases rapidly with increasing temperature, producing an extremely hard, adherent, insoluble coating on boiler tubes, especially in locations where heat flux is high. The compositions of several scales containing aluminum, magnesium, calcium, and silicate have been given above. Analcite and acmite, which form at high temperature, are invariably found beneath sludges of hydroxyapatite or serpentine, or under porous deposits of iron oxides. Occasionally, other extremely insoluble iron or magnesium silicates are also encountered, and now and then α-quartz, SiO_2, appears, usually originating from colloidal silica, finely divided silt, or sand in the feed water.

The precipitate formed by calcium in phosphate-treated boiler water is usually represented as the normal phosphate, $Ca_3(PO_4)_2$. Attempts to precipitate this salt in the laboratory, however, invariably produce hydroxyapatite, the formula of which can be written in various ways including $Ca_{10}(PO_4)_6(OH)_2$, $3Ca_3(PO_4)_2 \cdot Ca(OH)_2$, and $Ca_5(PO_4)_3OH$. A consideration of the solubility products $(Ca^{++})(CO_3^{--}) = 4.8 \times 10^{-9}$, $(Ca^{++})^3(PO_4^{---})^2 = 1.3 \times 10^{-32}$, and $(Ca^{++})^5(PO_4^{---})^3(OH^{-}) = 3 \times 10^{-58}$, indicates that the basic salt forms in boiler water. Magnesium forms similar salts such as $Mg_3(PO_4) \cdot Mg(OH)_2$ and $Mg_5(PO_4)_3OH$, which are undoubtedly much less soluble than $Mg(OH)_2$.

Latimer[1] notes that in aqueous solution silicic acids participate in complex equilibria on account of the small differences in free energy between the numerous polyacids and their salts. Thus, a variety of mineral types have been identified in boiler scales and deposits. Acmite, $Na_2O \cdot Fe_2O_3 \cdot 4SiO_2$, and analcite, $Na_2O \cdot Al_2O_3 \cdot 4SiO_2$, both of which are quite insoluble, require high temperature to form and, thus, are seldom found in boilers operated at less than 300 psi. On the other hand, serpentine, $3MgO \cdot SiO_2 \cdot 2H_2O$, and hydrated magnesium orthodisilicate, $Mg_3Si_2O_7 \cdot 2H_2O$, can form at any pressure. Garrels and Christ[2] show that magnetite, Fe_3O_4, is unstable relative to ferrous metasilicate, $FeSiO_3$, so that if an aqueous solution contains sufficient silica to satisfy all of the iron present, ferrous metasilicate forms in preference to magnetite. Iron silicate scales are exceedingly insoluble and adherent. Table I-1 summarizes the components of deposits most commonly found in boilers and related equipment.

In addition to the inorganic components mentioned, boiler deposits may also contain organic material including starch, quebracho, pyrogallol, sodium alginate, sodium mannuronate, chestnut tannin, and carbonized sulfonated lignins, all of which have been used as sludge conditioners. When oil is present as a contaminant in boiler water, loose scales may form, particularly in water wall tubes. Oil serves as a nucleus and binder for scaling at hot spots, although these scales are often merely baked mud that is easily dislodged by hammering the tubes. The "oil balls" found in steam drums and water wall headers are typical formations in turbulent sections; they are especially common in steam drums, where they are formed by the rolling motion of the water.

A large proportion of the fouling that occurs in recirculating cooling systems is occasioned by mishaps related to pH control and contamination of the water by reducing agents. Failures in some part of the acid injection system are the source of most acid contamination. When the pH falls below 5.5 corrosion inhibitor films are damaged; at about pH 4 hydrogen begins to evolve, and iron dissolves rapidly.

$$Fe + 2H^+ = Fe^{++} + H_2 \tag{I-2}$$

In a system treated with chromate, ferrous ion formed in Eq. (I-2) is oxidized to ferric ion.

TABLE I-1

Components of Boiler Deposits

Mineral	*Formula*	*Nature of Deposit*	*Usual Location and Form*
Acmite	$Na_2O \cdot Fe_2O_3 \cdot 4SiO_2$	Hard, adherent	Tube scale under hydroxyapatite or serpentine
Alpha quartz	SiO_2	Hard, adherent	Turbine blades, mud drum, tube scale
Amphibole	$MgO \cdot SiO_2$	Adherent binder	Tube scale and sludge
Analcite	$Na_2O \cdot Al_2O_3 \cdot 4SiO_2 \cdot 2H_2O$	Hard, adherent	Tube scale under hydroxyapatite or serpentine
Anhydrite	$CaSO_4$	Hard, adherent	Tube scale, generating tubes
Aragonite	$CaCO_3$	Hard, adherent	Tube scale, feed lines, sludge
Brucite	$Mg(OH)_2$	Flocculent	Sludge in mud drum and water wall headers
Copper	Cu	Electroplated layer	Boiler tubes and turbine blades
Cuprite	Cu_2O	Adherent layer	Turbine blades, boiler deposits
Gypsum	$CaSO_4 \cdot 2H_2O$	Hard, adherent	Tube scale, generating tubes
Hematite	Fe_2O_3	Binder	Throughout boiler
Hydroxyapatite	$Ca_{10}(PO_4)_6(OH)_2$	Flocculent	Mud drum, water walls, sludge
Magnesium phosphate	$Mg_3(PO_4)_2$	Adherent binder	Tubes, mud drum, water walls
Magnetite	Fe_3O_4	Protective film	All internal surfaces
Noselite	$4Na_2O \cdot 3Al_2O_3 \cdot 6SiO_2 \cdot SO_4$	Hard, adherent	Tube scale
Pectolite	$Na_2O \cdot 4CaO \cdot 6SiO_2 \cdot H_2O$	Hard, adherent	Tube scale
Serpentine	$3MgO \cdot 2SiO_2 \cdot H_2O$	Flocculent	Sludge
Sodalite	$3Na_2O \cdot 3Al_2O_3 \cdot 6SiO_2 \cdot 2NaCl$	Hard, adherent	Tube scale
Xonotlite	$5CaO \cdot 5SiO_2 \cdot H_2O$	Hard, adherent	Tube scale

$$3Fe^{++} + CrO_4^{--} + 8H^+ = 3Fe^{+++} + Cr^{+++} + 4H_2O \tag{I-3}$$

Above a pH of about 3 ferric ion is hydrolyzed to the hydrous oxide.

$$Fe^{+++} + 3H_2O = Fe(OH)_3 + 3H^+ \tag{I-4}$$

If caustic soda is now added, as it often is in an acid spill, excess ferrous ion and chromic ions in the system precipitate as the hydrous oxides when the pH reaches about 7.

$$Fe^{++} + 2OH^- = Fe(OH)_2 \tag{I-5}$$

$$Cr^{+++} + 3OH^- = Cr(OH)_3 \tag{I-6}$$

Ferrous hydroxide is then oxidized rapidly by oxygen in the water to ferric hydroxide, producing heavy, slimy fouling.

$$2Fe(OH)_2 + \tfrac{1}{2}O_2 + H_2O = 2Fe(OH)_3 \tag{I-7}$$

The voluminous and insoluble hydrous oxides settle in places where the flow of water is slow, then under-deposit corrosion ensues, adding further to the mass of fouling.

When reducing agents contaminate chromate-treated cooling water chromate is reduced to chromic hydroxide and corrosion rates increase rapidly. Two especially troublesome reducing agents are hydrogen sulfide and sulfur dioxide, both of which are common in chemical processing and petroleum refining. Catalytic cracking, for example, produces hydrogen sulfide and hydrogen cyanide from sulfur and nitrogen compounds in petroleum feed stocks. These acidic gases are evolved in process gas (methane and ethane). Admiralty brass exchangers used to cool the mixed gases are occasionally penetrated, allowing hydrogen sulfide and hydrogen cyanide to leak into the cooling water.

Hydrogen cyanide reacts with any ferrous ion in the water to form ferrocyanide [hexacyanoferrate (II) ion].

$$6HCN + Fe^{++} = Fe(CN)_6^{-4} + 6H^+ \tag{I-8}$$

Hydrogen sulfide is oxidized to sulfur by chromate ion, which is itself reduced to chromic hydroxide.

$$3H_2S + 2CrO_4^{--} + 2H_2O = 3S + 2Cr(OH)_3 + 4OH^- \tag{I-9}$$

If, as is usual, the chromate treatment also contains zinc ion, it is precipitated as ZnS, destroying the cathodic corrosion barrier. When an admiralty brass cooler is penetrated by hydrogen sulfide a tough, gelatinous sludge of ZnS, CuS, $Cr(OH)_3$, and elemental sulfur accumulates in the tubes and on the tube sheets.

Sulfur dioxide from stacks, SO_2-treating, and other sources occasionally is drawn into cooling towers. In addition to reducing chromate ion, the pH of the water is lowered by the hydrolysis of the gas.

$$SO_2 + H_2O = H^+ + HSO_3^- \qquad (I\text{-}10)$$

$$3SO_2 + 2CrO_4^{--} + 4H_2O = 3SO_4^{--} + 2Cr(OH)_3 + 2H^+ \qquad (I\text{-}11)$$

Process leaks occur frequently in petroleum refining that contaminate cooling water. Usually the bulk of a heavier oil collects on the surface of the water in the basin, from which it can easily be skimmed into a vacuum truck. Compounded lubricating oils, however, from leaking oil coolers, form extremely tight, thick emulsions of water, dirt, dispersant, and sometimes calcium salts; these emulsions tend to plug tubes and foul tube sheets of heat exchangers. Leaking oil seldom reduces chromate, but it coats the filling material in the cooling tower, particularly if it contains film packing, and interferes with the air/water contact.

Wind-borne dust is blown into cooling towers, dispersed in the water, then carried along until it reaches a place where the velocity of the water slows enough for the solids to settle and build up a deposit. This is especially troublesome during land filling, road work, or excavation, but every dry, windy day adds a great deal of dirt to a cooling system. Much of this settles in the basin of the cooling tower, but enough is carried into the cooling system to cause difficulties. Clays, for instance, are so finely divided that they readily form colloidal solutions, or sols, in cooling water that under certain circumstances undergo gelation, setting to a more-or-less homogeneous, firm, elastic gel containing 2–3 percent clay with the remainder water.(3) Particles of clay also tend to adhere to each other when wet, and deposit as impervious, sticky masses. Oil leaks further complicate matters, as clay and other soils have a marked affinity for oil.

If polyacrylamides are used to disperse mud through a cooling system, it is possible for soft, gelatinous masses to form if the polymer is added too rapidly. When this material enters the tubes of a heat exchanger, the heat transforms it to a hard, plastic mass that is most difficult to remove. Mechanical blockage also occurs frequently when wood fibers from cooling tower deterioration or chips from carpentry repairs lodge in heat exchanger tubes or upon the faces of tube sheets.

Calcium, magnesium, and ferric phosphate are commonly found in scales and deposits in phosphate-treated cooling systems. A particularly voluminous and insoluble salt is formed in cooling water treated with, or contaminated by, ferrocyanide. This is the familiar, intensely blue ferric ferrocyanide, $Fe_4[Fe(CN)_6]_3$, also called Prussian blue ($K_{SP} = 3.3 \times 10^{-41}$). Other organic salts that occur in water-formed deposits include calcium carbonate, calcium sulfate, and magnesium silicate.

I.1.b Corrosion Deposits

Corrosion is the result of the oxidation of metal by some oxidizing agent in the environment. The area over which the metal is oxidized is called the anode and that at which the oxidizing agent is reduced is called the cathode. These areas are separated, but usually are not far apart. As corrosion proceeds, electrons flow through the metal between these areas and ions flow from one area to the other through the solution. This system constitutes an electrochemical cell.

The oxidation (corrosion) of iron by hydrogen ion is represented by the following equation:

$$2H^+ + Fe = Fe^{++} + H_2 \qquad \text{(I-12)}$$

This reaction is opposed by an irreversible potential called the hydrogen overvoltage, which is markedly affected by the condition of the surface of the metal. Agents that raise the overvoltage oppose or inhibit corrosion, while conditions that facilitate the escape of hydrogen from the cathodic surface (roughness, debris, etc.) augment corrosion. Table I-2 lists several anodic–cathodic couples that can occur in aqeuous environments, for example, cooling systems or boilers.

Much iron corrosion is caused by failure to clean and passivate new equipment properly, and also by allowing metallic surfaces to become scaled or fouled. In general, the corrosion product of mild steel is ferric oxide, Fe_2O_3. In boilers under sulfite or hydrazine treatment magnetite, Fe_3O_4, is the usual product.

$$3Fe_2O_3 + SO_3^{--} = 2Fe_3O_4 + SO_4^{--} \qquad \text{(I-13)}$$

Oxides of iron are often mixed with other inorganic compounds, oil, tar,

TABLE I-2

Anodic–Cathodic Couples

Anode	*Cathode*
Stressed metal	Unstressed metal
Low oxygen concentration	High oxygen concentration
Clean metal	Fouled metal
Iron	Copper

wax, or coke and, if old, are likely to be quite insoluble. Copper and cuprous oxide derived from corrosion of copper alloys in preboiler equipment are also frequently seen in boiler corrosion deposits. Cupric ions dissolved from stage heaters, condensers, and evaporators subsequently plate on steel boiler tubes, become a cathode, and the surrounding and underlying steel corrodes.

Hydrogen sulfide is a source of corrosion in petroleum refineries, where reflux drums, stripping columns, and heat exchangers fabricated of admiralty brass, monel, and cupro-nickel alloys are exposed to it. Corrosion deposits from these units may contain sulfides of copper, nickel, zinc, and iron. Regenerators in hydrofluoric alkylation plants, which are also made of monel, are often found to contain nickel, copper, and iron fluorides as products of corrosion.

I.1.c Microbiological Deposits

Water in an open recirculating cooling system is continuously infested with a variety of nuisance microorganisms indigenous to soil including bacteria, algae, and fungi.[4] These organisms are introduced in wind-borne dirt and once in the water they proliferate unless checked by the addition of microbicides. Algae are often visible as green, felt-like mats in internal sections of cooling towers that are wet and accessible to sunlight. Filamentous and capsulated colonial algae coat splash packing and interior structural members, thereby interfering with the formation of droplets that is essential for intimate contact between water and air; they are otherwise innocuous. As these microorganisms require sunlight to live, they are not found within heat transfer equipment. Although in the field of water treatment there is frequent reference to "dead algae," as a practical matter, there is no such thing. In common with other microorganisms, algae when they expire, immediately undergo autolysis and shortly vanish; when exposed to a microbicide they dry up and disappear.

Of considerably more concern are soil bacteria that secrete slime. These microorganisms are surrounded by a capsule, which if diffuse and gelatinous, is called a slime layer. Its composition is variable, but in the main it is a secretion of polysaccharide that arises in the cytoplasmic membrane. Slime is a normal product of metabolism, and is apparently composed of degradation products of the cell wall. *Aerobacter* and *Pseudomonas* are the predominant bacterial genera in cooling systems. Both of these elaborate gelatinous capsules which are sticky, and thus can accumulate mud and other debris into masses that clog screens and impede circulation in heat exchangers and surface condensers. *Aerobacter* produces a more extensive capsule than does *Pseudomonas*, but profuse slime is particularly associated with the latter, which generates material ranging in consistency from that of molasses to rubber. Microbiological growths, in general, possess three characteristics that can be used to identify them: they are stringy; they are sticky; and they smell bad. As microorganisms are composed of 70–80 percent water, dehydrated bacterial films on heat exchanger tubes take the form of dry leaves that are easily flushed out.

In manufacturing facilities so situated that they can use sea water for cooling, heat transfer equipment is subject to seasonal plugging by various marine bivalve molluscs including mussels, scallops, and Asiatic clams; barnacles also are sometimes troublesome. Immature shellfish enter the salt water distribution systems, then grow within these systems. Chlorine is customarily injected at pumping stations to control marine growth, but unless applied continuously it is ineffective, as the animals close their shells while slugs of chlorine pass, and are unharmed. Eventually, they collect and block tube sheets or obstruct water flow in the shells of heat exchangers. This kind of fouling, of course, must be mechanically removed.

I.1.d Petroleum Process Fouling

In petroleum refining, fuel is conserved by using hot process streams to preheat feed stocks. Thus, crude petroleum fed to the atmospheric distillation column of a crude unit successively exchanges heat with the overhead and various side cuts before entering the atmospheric furnace. The crude feed usually passes through the tubes and the distillates through the shells of the heat exchangers. The presence of miscellaneous resinous, asphaltic, and unstable sulfur compounds leads to fouling of tubes in the crude preheat train, which decreases the rate of heat transfer and increases the pressure drop across the exchangers. This in turn increases the amount of fuel necessary to fire the furnace and eventually decreases the production rate.

Fouling in these exchangers consists of black, tarry masses of petroleum-derived material, often intimately mixed with ferrous sulfide, iron oxides, calcium sulfate, dirt, and occasionally, elemental sulfur. These matrices of organic and inorganic materials may be very difficult to remove completely. When coke accumulates in furnace tubes it is removed by steam-air decoking, in which process steam containing enough air to provide about 1 percent of oxygen is passed through the heated furnace tubes, generating a burn wave that can be easily followed from outside the furnace.

Two unusual polynuclear aromatic compounds resulting from the hydrocracking of tars are encountered in deposits from these plants. Coronene and ovalene are composed of six-membered rings condensed in the configuration of chicken wire.[5] The former is so called because its structural formula is a symmetrical corona, or crown, of six fused benzene rings.

Coronene

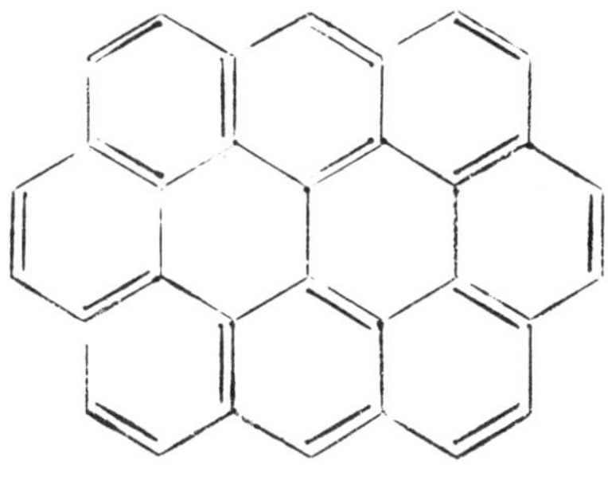

Ovalene

Coronene, when pure, forms light yellow needles having a melting point of 440°C (824°F); ovalene forms light orange needles having a melting point of 473°C (883°F). These high-melting compounds are difficult to remove when deposited in hydrocrackers. Other polycyclic aromatic hydrocarbons that have been identified in such deposits include alkylcoronene, phenanthrene, pyrene, benzopyrene, benzoperylene, and benzobisanthene.[5a]

I.1.e Furnace Deposits

Furnaces in which fuel oil is burned are subject to corrosion and the formation of deposits arising from the presence of certain chemical elements in the fuel oil. The most prevalent of these are listed in Table I-3.

TABLE I-3

Metallic Contaminants in Fuel Oil

Element	*Concentration range,* ppm	*Form or source*
Aluminum	0–25	Clay, dirt, catalyst
Calcium	0–190	Emulsified salts
Iron	0–250	Porphyrins, rust
Magnesium	0–20	Emulsified salts
Nickel	0–230	Porphyrins, corrosion products
Silicon	0–275	Clay, dirt
Sodium	5–250	Emulsified salts
Vanadium	2–300	Porphyrins

Iron, nickel, and vanadium usually predominate, the latter two ordinarily comprising more than 10 percent of the total ash. Vanadium, nickel, and iron may be present in oil-soluble porphyrin-metal chelated complexes and probably other complexes with nitrogen compounds and soaps. As the latter are effective emulsifying agents, they bind water or brine in extremely stable emulsions. These are especially troublesome on ships where fuel oil bunkers are customarily ballasted with sea water. It is not uncommon to find that the fuel oil from a ballasted bunker contains half a percent of emulsified sea water. In addition to oil-soluble compounds in crude oil, scale, refining catalysts, corrosion products, and dirt are also sources of metallic contaminants in fuel oil.

Nonvolatile vanadium, which concentrates in the residuum from which heavy fuel oil is produced, is intensely corrosive to refractories used in furnaces.[6] It is most destructive to fire-clay bricks; alumina and magnesia firebricks are much less affected. Vanadium oxides form low-melting eutectics with the clay, forming hard, glassy slags that are very difficult to remove. Vanadium pentoxide is an active and corrosive flux, and its low melting point (1216°F), combined with its tendency to form low-melting eutectic mixtures with other compounds—notably sodium salts—contributes to the formation of slag that coats tubes, superheaters, and supporting hardware when fuel oils that contain vanadium are burned.

Slag accumulates when metallic salts and oxides vaporized in the furnace condense on cooler surfaces within the furnace, bridging tubes, blocking gas passages, and reducing heat transfer. Sodium salts lower the melting points of vanadium oxides, particles of which pass into the furnace as a sticky sublimate. Sodium chloride, sodium sulfate, nickel salts, and complex vanadates all precipitate from the vapor state, condensing as liquid droplets in the radiant and superheater sections of the furnace. Molten vanadates and particles of ash that are semimolten, or at the stage of incipient sintering, then deposit on various surfaces as a dense tenacious slag, which upon cooling is of a vitreous consistency. Other solid particles including calcium, magnesium, and iron oxides, as well as some silicates and fly ash are physically entrapped in the molten or sticky slag. Table I-4 lists names, formulas, and fusion temperatures of several compounds that have been identified in furnace slags, all of them having melting points well below flame temperatures in the radiant section, i.e., 2600-3000°F. They condense there first, of course, because the tube temperatures are 450-650°F, then throughout the convection section as the temperature of the flue gas falls in a couple of seconds from 3000°F to about 700°F.

TABLE I-4

Chemical Compounds in Furnace Deposits

Name	*Formula*	*Melting point* °F	°C
Sodium chloride	$NaCl$	1486	808
Sodium sulfate	Na_2SO_4	1623	884
Sodium vanadate	Na_3VO_4	1590	866
Sodium metavanadate	$Na_2O \cdot V_2O_5$	1165	630
Sodium divanadate	$Na_4V_2O_7$	1209	654
Nickel pyrovanadate	$2NiO \cdot V_2O_5$	1650	899
Vanadium pentoxide	V_2O_5	1216	658

The results of a chemical analysis of a slag from the convection passages of a 900-psi marine boiler furnace are shown in Table I-5.

TABLE I-5

Chemical Analysis of Furnace Slag

Component	*% by weight*
Vanadium oxide, V_2O_5	44
Nickel oxide, NiO	3
Iron oxide, Fe_2O_3	1
Sodium oxide, Na_2O	12
Calcium oxide, CaO	1
Sulfur trioxide, SO_3	35
Aluminum oxide, Al_2O_3	0.2
Magnesium oxide, MgO	0.4
Lead oxide, PbO	0.4
Silicon dioxide, SiO_2	0.6
Carbon dioxide, CO_2	2
Carbon, C	0.4

In general, the concentration of vanadium is highest in the radiant section, sulfates are highest in the convection section, and iron is highest in the cooler sections where corrosion is exacerbated by the condensation of sulfuric acid.

In addition to their insulating effects, vanadium slags attack steel, especially tube brackets, superheater hangers, and support bracelets, but much more virulent corrosion occurs in the economizer and air heater by sulfuric acid formed in the combustion of sulfur-containing fuel oils, in which the concentration of sulfur ranges from 0.2 to 6 percent. With minimum excess air the concentration of SO_3 in flue gas seldom exceeds 10 ppm, but in the presence of contact catalysts of the first transition series (iron, nickel, vanadium, etc.), its concentration can reach 20–100 ppm.

$$S + O_2 = SO_2 \tag{I-14}$$

$$SO_2 + \tfrac{1}{2}O_2 = SO_3 \tag{I-15}$$

$$SO_3 + H_2O = H_2SO_4 \tag{I-16}$$

$$H_2SO_4 + Fe = FeSO_4 + H_2 \tag{I-17}$$

When the temperature of the exiting stack gas is below the dew point of sulfuric acid—about 350°F—the economizer and air heater are severely attacked by a condensed film of concentrated sulfuric acid. Also, passages in the economizer are likely to be blocked by deposits of ferric sulfate and fly ash adhering to the damp surfaces.

I.2 MECHANISM OF CORROSION

Several products of corrosion have been cited in the previous section as components of fouling deposits. In order to remove these from industrial equipment it is in most instances necessary to resort to aggressive solvents, in particular acids, which are considerably more corrosive to the equipment than the materials normally processed in it. Obviously, steps must be taken to limit the attack on substrate metal while the dissolution of a scale or deposit is underway. This matter is discussed in Chapter 2, but here it is worthwhile considering several aspects of what may be called normal corrosion.

I.2.a Oxidation Potentials

It is common in metallurgical work to speak of one metal being noble to another. This means that of the two, the nobler metal has the greater tendency to retain its electrons. Thus, the more noble a metal the less its tendency to oxidize, or corrode. In Table I-6 several common metals and alloys are tabulated in increasing order of nobility, in which order any metal in the list is anodic to those below it. Thus, if aluminum were coupled to copper and dipped in a solution of some neutral electrolyte, aluminum would tend to dissolve, i.e., be corroded. In industrial practice, insulating gaskets are used to prevent electrochemical coupling of dissimilar metals.

TABLE I-6

Electrochemical Series of Selected Metals

Aluminum
Mild Steel (99Fe, 0.75Mn, 0.25C)
Stainless Steel (74Fe, 18Cr, 8 Ni, 0.18C)
Admiralty Brass (70Cu, 29Zn, 1Sn)
Copper
Cupro-nickel (70Cu, 30Ni)
Monel (60Ni, 33Cu, 6.5Fe)

In Table I-7 standard oxidation potentials are tabulated of several metals that are subject to corrosion in aqueous solutions.

TABLE I-7

Standard Oxidation Potentials in Various Solutions

Metal	I *Acid solution* pH = 0	II *Carbonate solution* pH = 11 (CO_3 = 1M)	III *Alkaline solution* pH = 14	IV *Sulfide solution* pH = 11.7 (S = 1M)
Aluminum	1.66	*	2.31	*
Zinc	0.76	1.06	1.24	1.44
Iron	0.44	0.76	0.88	0.97
Nickel	0.25	0.45	0.72	0.83
Copper	−0.34	−0.05	0.22	0.76

* Aluminum ion is completely hydrolyzed in carbonate and sulfide solutions.

The four typical half-reactions for which standard oxidation potentials are summarized in Table I-7 are:

$$\text{I}\quad Fe = Fe^{++} + 2e^- \qquad E^\circ = 0.44$$

$$\text{II}\quad Fe + CO_3^{-2} = FeCO_3 + 2e^- \qquad E^\circ = 0.76$$

$$\text{III}\quad Fe + 2OH^- = Fe(OH)_2 + 2e^- \qquad E^\circ = 0.88$$

$$\text{IV}\quad Fe + S^{-2} = FeS + 2e^- \qquad E^\circ = 0.97$$

In column I, which lists E° values in one molal hydrogen ion, the large value of 1.66 V indicates that aluminum can be oxidized easily to Al^{+3}, i.e., corroded. Similarly, in comparing the potentials in column I of iron and nickel, it is seen that nickelous ion should oxidize metallic iron to ferrous ion:

$$Fe = Fe^{++} + 2e^- \qquad E^\circ = 0.44$$

$$Ni^{++} + 2e^- = Ni \qquad E^\circ = -0.25$$

$$Fe + Ni^{++} = Fe^{++} + Ni \qquad E^\circ = 0.19\ V \tag{I-18}$$

A positive potential for the net reaction shows that the reaction proceeds spontaneously as printed. In general, a metal gives up its electrons to an ion whose affinity for electrons is greater than that of the metal. That is, the baser metal is oxidized by ions of the nobler. Table I-7 shows also that as pH increases, oxidation potentials become more positive, and that the highest potentials occur in sulfide solution (column IV). Thus, aluminum or zinc would not be expected to be satisfactory for handling alkaline solutions, and it can be futher inferred that all of the metals in Table I-7 are corroded by sulfide or hydrogen sulfide.

I.2.b Hydrogen Overvoltage and Free Energy

The standard oxidation potential of the hydrogen electrode, in which the concentration of hydrogen ion is 1 molal and the pressure of hydrogen gas is 1 atmosphere, is defined as 0.00 V. If this potential is compared with that of iron it is readily apparent that steel has a pronounced tendency to be oxidized by acidic solutions.

$$\begin{array}{ll} Fe = Fe^{++} + 2e^- & E^\circ = 0.44 \\ 2H^+ + 2e^- = H_2 & E^\circ = 0.00 \\ \hline Fe + 2H^+ = Fe^{++} + H_2 & E^\circ = 0.44\ V \end{array} \qquad \text{(I-19)}$$

The Nernst equation can be used to calculate the potential of the hydrogen electrode at concentrations of hydrogen ion differing from 1 molal.

$$E = E^\circ - \frac{RT}{nF} \log_e \frac{[\text{oxidized form}]}{[\text{reduced form}]} \qquad \text{(I-20)}$$

Where:

R = 8.314 absolute J/degree mole
T = 298.16°
n = number of electrons transferred (equivalents)
F = 96,484 absolute coulombs

Also,

$$\log_e x = 2.303 \log_{10} x$$

Thus,

$$E_H = 0.00 - \frac{2.303 \times 8.314 \times 298.16}{2 \times 96{,}484} \log_{10} [(H^+)^2/(p_{H_2})]$$

$$= 0.00 - 0.0592 \log_{10} [(H^+)/(p_{H_2})^{1/2}]$$

As it is convenient to measure small concentrations of hydrogen ion as pH, it is noted that $pH = -\log_{10}(H^+)$, therefore,

$$E_H = 0.0592\,pH + 0.0296 \log_{10} p_{H_2} \qquad \text{(I-21)}$$

If p_{H_2} is assumed to be 1 atm., the potential of the hydrogen electrode is given by:

$$E_H = 0.0592\,pH \qquad \text{(I-22)}$$

Thus, at pH 7 its potential is 0.414 V and the potential of the net reaction shown as Eq. (I-19) is (0.440 − 0.414) = 0.026 V. Although this potential is positive, the corrosion rate of iron in aerated water at pH 7 is relatively slow, and in fact the rate is essentially constant, at about 10 mpy, between pH 4 and pH 10, as the result of the phenomenon of overvoltage.

Hydrogen overvoltage is the excess voltage above the reversible potential that must be applied to reduce hydrogen ion to hydrogen gas. Values depend upon current density, the metal used, roughness, surface defects, and other factors. For iron at a current density of 10^{-4} amp/cm^2, Latimer(1) gives 0.22 V. Experimentally, at about pH 4, which corresponds to 0.237 V, hydrogen begins to evolve on an iron surface. Overvoltage decreases with increasing temperature and surface roughness, and with decreasing current density, all of which tend to facilitate the release of hydrogen bubbles from metal surfaces.

The free energy, ΔG, of a chemical reaction such as that represented by Eq. (I-19) can be calculated from the reaction potential as follows:

$$\Delta G = -nFE \qquad \text{(I-23)}$$

When iron reduces hydrogen ion to hydrogen gas under standard conditions, the free energy of the reaction is:

$$\Delta G = -2 \times 96{,}484 \times 0.44$$
$$= -84{,}906 \text{ J}$$
$$= -20{,}293 \text{ Cal}$$

Chemical free energy is essentially the work required to break chemical bonds, so if it does not decrease, chemical reaction cannot occur spontaneously. Thus, there is nothing to prevent writing Eq. (I-19) as:

$$H_2 + Fe^{++} = 2H^+ + Fe \qquad \text{(I-24)}$$

If this is done, however, then E° would be –0.44 V and ΔG would be +20,293 Cal, indicating an increase in free energy, which is thermodynamically impossible, unless external energy is supplied, as in electrolysis.

The reduction of hydrogen ion takes place in steps, the first of which is the capture of an electron and the adsorption of the resulting atom of hydrogen on the metal surface.

$$H^+ + e^- = H\,(ads) \qquad \text{(I-25)}$$

The second step may be either

$$2H\,(ads) = H_2\ (g) \qquad \text{(I-26)}$$

or

$$H^+ + e^- + H\,(ads) = H_2\ (g) \qquad \text{(I-27)}$$

The reaction in Eq. (I-25) is considered by West(7) to be the slowest of the three and, thus, the rate-determining step; Uhlig(8) suggests that the reaction in Eq. (I-27) plays that role in the reduction of hydrogen ion.

Iron also reduces hydrogen in highly alkaline environments such as those that lead to caustic gouging within boilers. West(7) proposes the following sequence of reactions:

$$Na^+ + e^- = Na \qquad \text{(I-28)}$$
$$Na + H_2O = Na^+ + OH^- + H\,(ads) \qquad \text{(I-29)}$$
$$2H\,(ads) = H_2\ (g) \qquad \text{(I-30)}$$

The net reaction with iron is then:

$$Fe + 2H_2O = Fe(OH)_2 + H_2\ (g) \qquad (I\text{-}31)$$

In the highly alkaline water where hydrogen evolution occurs, however, the other product is hypoferrite ion.

$$Fe + 2OH^- = FeO_2^{--} + H_2\ (g) \qquad (I\text{-}32)$$

that is,

$$Fe(OH)_2 + 2OH^- = FeO_2^{--} + 2H_2O \qquad (I\text{-}33)$$

I.2.c Polarization and the Corrosion Current

Fig. I.1 is a schematic representation of an electrochemical cell in which current flows as the result of a transfer of electrons during an oxidation-reduction reaction. In the process of cleaning steel with acids, the net half-reaction at the anode is the oxidation of iron to ferrous ion with the con-

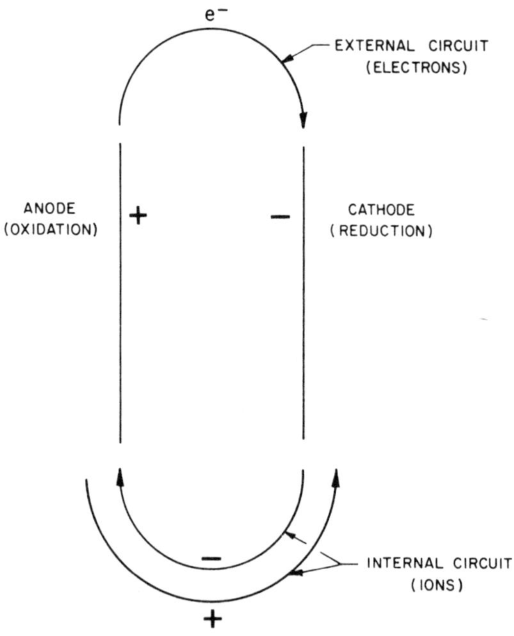

Fig. I.1 Diagram of an electrochemical cell

comitant reduction of hydrogen ion to hydrogen gas at the cathode. The flow of electrons through the external circuit is the corrosion current, which is limited by the rate at which electrons are accepted by the oxidizing agent (H^+) at the cathode. When current flows in any electrochemical cell, the open-circuit potential of the anode becomes more cathodic, while that of the cathode becomes more negative: that is, they become polarized. Polarization, in volts, is the difference between the open-circuit potential and the new potential established by a change in flow of current.

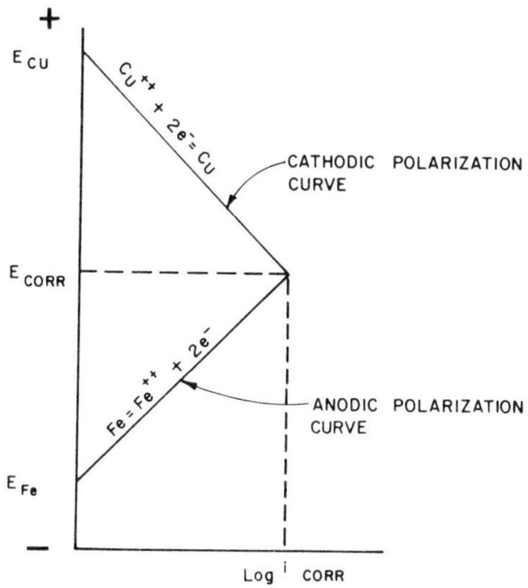

Fig. I.2 Polarization curves

The relationship between the corrosion, or exchange, current and potential is shown by plotting anodic and cathodic polarization curves[9] to produce an Evans[10] diagram. Fig. I.2 is the diagram of a polarized copper–iron galvanic cell. The slopes of the polarization curves are defined as the rate of change of potential with respect to the logarithm of the corresponding current; they are called Tafel[11] slopes.

$$dE/d(\log_{10} i) = \beta \quad \text{(I-34)}$$

Limiting the flow of the corrosion current is an eminently important consideration in devising methods for chemical cleaning. Stern[12] has published an article reviewing the fundamentals of electrode processes in corrosion that the interested general reader will find instructive.

I.2.d Surface Effects in Corrosion

Localized corrosion is caused by potential differences between discrete areas on a metal surface. Minor defects in that surface, including crystallographic vacancies, dislocations in the crystal lattice, or a disorder among the atoms, give rise to unusually high lattice energy that makes these locations a few millivolts anodic to the surrounding metal. In the presence of an oxidizing agent to accept electrons, metal ions go into solution in these localized regions, while electrons flow through the metal to cathodic areas where the oxidizing agent is reduced. Similarly, a plating of metallic copper on a steel surface enhances the corrosion of iron; in effect, the less noble metal gives cathodic protection to the more noble in this bimetallic galvanic cell. Another example is cementite, Fe_3C, a component of high-carbon steel, which behaves as a cathode on steel surfaces. Finally, grain boundaries always contain defects such as crystallographic mismatches.

In aerated solutions the cathodic half-reactions are usually either

$$2H^+ + 2e^- = H_2 \qquad E^\circ = 0.00 \qquad \text{(I-35)}$$

or

$$2H^+ + \tfrac{1}{2}O_2 + 2e^- = H_2O \qquad E^\circ = 1.23 \qquad \text{(I-36)}$$

In neutral or alkaline solutions it is

$$H_2O + \tfrac{1}{2}O_2 + 2e^- = 2OH^- \qquad E^\circ = 0.81\ (\text{pH } 7) \qquad \text{(I-37)}$$

$$E^\circ = 0.40\ (\text{pH } 14)$$

Although the oxidation potentials of oxygen are significantly higher than that of hydrogen ion, corrosion rates are generally lower with oxygen, primarily because its solubility in water is small, and its rate of reduction is diffusion limited. In addition, the rate of diffusion of oxygen is much lower than that of hydrogen ion.

A certain degree of natural protection is afforded by oxide films that form on the surfaces of metals exposed to aerated water. Certain so-called aggressive anions, however, notably chloride, are electrostatically attracted to anodic defects, where they concentrate and promote the dissolution of the protective film. Chloride ions damage oxide films by forming complexes with the metal ions, by displacing hydroxyl ion from the surface of the metal, and by lowering the hydrogen overvoltage of the surface. West[7] also points out that as the film attempts to repair itself, hydroxyl ions are consumed so that the concentration of hydrogen ion increases within the defect, or pore, reaching a pH of about 2, even if the bulk solution is neutral. This type of action produces pitting, especially at anodes that are small relative to the surrounding cathodic area.

Similar effects are caused by mill scale formed at high temperature on ferritic surfaces. This scale is brittle with a pronounced tendency to spall. As it always contains defects and microfissures, it is essential that it be removed completely before putting a new piece of equipment, such as a boiler or heat exchanger, into aqueous service.

I.2.e Measurement of Corrosion Rates

Methods for estimating corrosion rates in chemical cleaning solutions are based upon weight losses of metal specimens (coupons, polished nipples, tubes, or rotating rods), and electrical measurements, either linear polarization or resistance. Coupons are weighed strips of metal that are either statically immersed or else suspended in a suitable holder in a flowing stream of cleaning solution for some length of time, usually 6 h in laboratory tests, after which they are removed, cleaned, and reweighed. As the method lacks precision under the most favorable conditions, it is important that all tests be carried out the same way with respect to the preparation of the coupons, duration of exposure, and rate of solution flow. The American Society for Testing and Materials[13] has published a standard procedure giving exhaustive details for cutting and preparing coupons of various metals for determining the corrosivity of cleaning solvents.

It is important that specimens be cut by sawing or milling rather than by shearing, as heavy pitting always takes place on sheared edges or otherwise stressed locations. Directions are provided in the ASTM method cited for measuring corrosivity of inhibited chemical cleaning solutions to statically immersed specimens and to those installed in a stream of solution flowing at 2 fps. The latter can be approximated by rotating a rod at appropriate speed in the solution to be tested. Bregman and Newman[14] have also described the preparation and use of rotating rods for comparative laboratory corrosion tests. The method also contains procedures for measuring corrosivity in the presence of selected cations, including aluminum, ferric, and the alkaline earths, as well as anions such as phosphate, sulfate, and silicate. The effects of coatings of magnetite or synthetic deposits on corrosivity of a solvent to metal can also be tested, if desired. Nonferrous metals are exposed for 6 h to cleaning solvents at 100–150°F; ferrous metals are checked at 150–180°F. The weight loss is reported as mg/6 h.

A further complication in coupon testing is crevice corrosion. This mode of attack, caused by the existence of oxygen concentration cells, occurs in places where contact is made between metal and nonmetal, in threaded pipe connections, in spaces between tube and tubesheet in heat exchangers, or in places where two pieces of metal are fastened together in such a way that chemical solvent can penetrate between them.

Another important influence on corrosion rates during chemical cleaning is the linear velocity of the cleaning solvent; it should flow through heat transfer equipment and piping at no more than 1–2 fps to avoid the erosion of metallic surfaces. At flow rates greater than about 4 fps the filming amines commonly used as corrosion inhibitors in acidic cleaning solvents are stripped from the metal substrate, which is then exposed to attack by hydrogen ion. Coupons installed in industrial equipment, especially steam generators, to check corrosion during chemical cleaning must be carefully placed. If they are mounted near the filling line on a boiler, for instance, filling, draining, and positive circulation of acid cause turbulence around the coupons that greatly increases their apparent corrosion rate.

For many years the rate of corrosion of metals has been expressed in mils per year, mpy, of penetration, but with the adoption of the metric system the preferred units are milligrams per square decimeter per day, mdd. As both of these terms as well as several others are still in use, the methods of calculating several expressions are given here. Table I-8 contains the densities of several metals and alloys that are required for calculating mpy.

The average corrosion rate, in mils per year, is calculated from the weight loss of a coupon as follows:

$$\text{Average penetration, mpy} = \frac{w \times 365 \times 10^3}{(2.54)^3 A d t} \tag{I-38}$$

TABLE I-8

Densities of Common Metals

Metal or alloy	*Density, d,* g/cm^3
Admiralty brass	8.17
Cupro-nickel, 70-30	8.50
Monel metal	8.85
Steel, mild	7.86
Steel, stainless, No. 304	7.87
Steel, stainless, No. 410	7.69
Titanium	4.51

where:

w = weight loss, g
A = surface area of coupon, $in.^2$
d = density of metal, g/cm^3
t = time, days

Other terms for expressing corrosion rates include milligrams per square decimeter per day, mdd; milligrams per square centimeter per day, mcd; and pounds per square foot per day, lb/ft^2-day. These are calculated as follows:

$$\text{mdd} = \frac{w(\text{mg}) \times 24(\text{h/day})}{(\text{dm}^2)/(3.94)^2(\text{in.}^2) \times A(\text{in.}^2) \times t(\text{h})} \qquad \text{(I-39)}$$

$$\text{mcd} = \frac{w(\text{mg}) \times 24(\text{h/day})}{(2.54)^2(\text{cm}^2)/(\text{in.}^2) \times A(\text{in.}^2) \times t(\text{h})} \qquad \text{(I-40)}$$

$$\text{lb/ft}^2\text{-h} = \frac{(\text{g})}{454(\text{g/lb}) \times A(\text{in.}^2) \times 0.00694(\text{ft}^2/\text{in.}^2) \times t(\text{h})} \qquad \text{(I-41)}$$

It is often convenient to convert from one expression of corrosion rate to another. This is best done by a dimensional analysis. The expression in Eq. (I-39), for example, can be converted into that of Eq. (I-38) as follows:

$$\frac{\text{mg} \times 24(\text{h/day})}{\text{dm}^2/(3.94)^2(\text{cm}^2/\text{in.}^2) \times \text{in.}^2 \times \text{h}} \times \frac{365(\text{days/yr}) \times 394(\text{mils/cm})}{(2.54)^2(\text{cm}^2/\text{in.}^2) \times 10^3(\text{mg/g}) \times 15.5(\text{in.}^2/\text{dm}^2) \times d\,(\text{g/cm}^3)}$$

or,

$$\text{mdd} \times 1.44/d = \text{mpy}$$

Other conversions are:

$$\begin{aligned} \text{mdd} \times 7.5 \times 10^{-3} &= \text{lb/ft}^2\text{-yr} \\ \text{mdd} \times 10^{-2} &= \text{mcd} \\ \text{mdd} \times 1.6 \times 10^{-4}/d &= \text{mph} \\ \text{mdd} \times 2.7 \times 10^{-6} &= \text{g/in.}^2\text{-h} \end{aligned}$$

Freedman, et al.[(15)] have described a device for measuring electrical resistance, which they used to monitor corrosion rates. The principle of the measurement is that the resistance of an electrical conductor is inversely proportional to its cross-sectional area. The corrosivity of a cleaning solvent can be measured by inserting a metallic sensor into a pipe or header carrying the solvent. As corrosion proceeds the cross section of the metal specimen decreases and its resistance increases, the change in the latter being correlated with corrosivity.

A much more useful measurement in aqueous solvents is that of linear polarization, or polarization resistance. As explained in Section I.2.c, when current flows in an electrochemical cell the anode becomes more cathodic and the cathode becomes more anodic than the open-circuit potentials of each electrode. The extent of the decrease in potential at each electrode caused by the net current to or from an electrode is called the polarization. The electrode potential of a metal freely corroding in an aqueous solution is defined by the intersection of the polarization curves of the anode and cathode, which are obtained by plotting the potential against the logarithm of the current flowing between the two electrodes. The point of intersection determines the potential of the corroding metal and also the corrosion current. The less the slope of these curves, the higher the corrosion rate. Commercial meters are available that register the slope of the apparent polarization curve in the region near the potential of freely corroding metal, which is proportional to the rate of corrosion. Marsh and Schaschl[(16)] have investigated the use of these instantaneous corrosion rates for evaluating corrosion inhibitors. A standard procedure[17] is available for measuring corrosivity by either linear polarization or resistance. These electrical devices have the same limitations as metal coupons in that they give values that represent the rate of corrosion of the sensor at the point of insertion. Generally, results by electrical measurements are higher than those obtained by weighing coupons.

I.3 HAZARDS OF CHEMICAL CLEANING

There are several sources of potential injury to personnel engaged in chemical cleaning to be considered in this section. Among these hazards are the chemical cleaning solutions themselves, toxic gases generated by reactions between solvents and fouling deposits, pressure buildup from the generation of hydrogen or the release of carbon dioxide from carbonates, flammable substances including hydrogen and hydrocarbons, and the possible presence of pyrophoric materials. In addition, there are many opportunities for metallurgical damage; the latter, however, are discussed in Chapter II.

I.3.a Chemical Cleaning Agents

The majority of chemical cleanings are done with dilute mineral acids, most often hydrochloric acid. Less frequently, sulfuric and phosphoric acids are used, usually because of metallurgical considerations. All of these reagents produce burns when concentrated, particularly sulfuric acid, which is highly corrosive to living tissues. It is a powerful dehydrating agent that causes severe burns and extensive damage. Because of the very large heat of solution of sulfuric acid—about 18,000 calories per mole—the acid should invariably be added to water rather than the reverse. Chromic acid (CrO_3) is sometimes used to oxidize cokes or tarry fouling in petroleum refining and chemical processing equipment. The concentrated acid must be judiciously handled, however, as it can cause a fire or violent oxidation if brought into contact with combustible material. The chemical causes irritation and ulceration of the skin and, if ingested, severely irritates the gastro-intestinal tract. Sulfamic acid, which is seldom used now, is moderately irritating to the skin and mucosa.

Among the organic acids used for chemical cleaning, oxalic and formic acids are corrosive to the skin, the latter being especially dangerous. Oxalic acid, if taken internally, produces vomiting, diarrhea, and renal damage.

Citric and gluconic acids are relatively innocuous, but hydroxyacetic acid, often recommended for cleaning utility boilers, is irritating to the skin and mucosa.

Ammonium bifluoride, NH_4HF_2, is commonly used in combination with hydrochloric acid to solubilize silicate scales in boilers and heat exchangers. This combination should be handled as hydrofluoric acid. It causes painful skin lesions and, if ingested, nausea, vomiting, diarrhea, convulsions, circulatory collapse, and death may ensue. Other hazardous cleaning materials include sodium hydroxide, thiourea, chlorinated hydrocarbons, and sodium nitrite. The latter, if inadvertently acidified, releases the toxic brown gas, NO_2, which rapidly causes severe damage to the lungs.

$$2H^+ + 2NO_2^- = NO + NO_2 + H_2O \qquad (I\text{-}42)$$

It is imperative that people engaged in chemical cleaning be provided with (and use) adequate safety equipment such as chemical goggles, protective clothing, rubber gloves, boots, aprons, and face shields. The eyes, especially, must be diligently protected from all cleaning agents. Also, safety showers, either portable or permanent, are essential near the site of a chemical cleaning operation.

I.3.b Toxic Gases

The generation of toxic gases is a special hazard encountered in cleaning petroleum refining equipment. Hydrogen sulfide, H_2S, is often liberated during the cleaning of vaporizer furnace tubes and feed preheat exchangers in crude petroleum service. Hydrogen sulfide is an extremely hazardous, flammable, toxic gas with an odor resembling that of rotten eggs. Headache, nausea, and dizziness occur after moderate exposure; in higher concentrations collapse, coma, and death may occur within a few seconds. When this hazard exists, alkaline scrubbers are installed to absorb the gas, and workers must be supplied with self-contained breathing apparatus. In addition, a suitable test kit* for measuring H_2S in the atmosphere should be provided; it is also common practice to suspend strips of lead acetate test paper throughout the chemical cleaning area. These strips should be moistened, for maximum sensitivity, but they are used only as a

* Either the MSA Universal Testing Kit, Mine Safety Appliances Company, 400 Penn Center Boulevard, Pittsburgh, PA 15235, or the Dräger Multi Gas Detector, Drägerwerk AG Lübeck, Postfach 13 39, Moislinger Alee 53/55, D-2400 Lübeck 1, Federal Republic of Germany, is satisfactory.

qualitative indicator supplemental to a quantitative monitoring device.

Scales and deposits in the gas recovery sections of catalytic cracking units often release both hydrogen cyanide, HCN, and hydrogen sulfide when exposed to acids. Furthermore, in the presence of cupric ion, cyanogen, C_2N_2, may be generated.

$$2Cu^{++} + 4CN^- = C_2N_2 + 2CuCN \qquad \text{(I-43)}$$

Both hydrogen cyanide and cyanogen have a characteristic almond odor (that of cyanogen is the more pungent) and they are intensely poisonous. High concentrations produce paralysis, convulsions, and respiratory failure; lesser concentrations cause headache, dizziness, nausea, and vomiting. Cyanogen specifically contracts eye muscles causing the eyes to protrude painfully; in an alkaline scrubber it is rapidly hydrolyzed to cyanide and cyanate.

$$C_2N_2 + 2OH^- = CN^- + CNO^- + H_2O \qquad \text{(I-44)}$$

Alpha-brasses, which are solid solutions of up to 40 percent zinc in copper, are subject to dezincification when exposed to sea water or soft fresh water. In this corrosion process the more reactive component (zinc) is selectively oxidized leaving the metallic surface enriched in the less reactive component (copper). The corrosion rate of zinc can be slowed by adding a small amount of arsenic, which has a high hydrogen overvoltage, to the alloy. Arsenite has been used as a corrosion inhibitor in monoethanolamine gas purification systems. When alloys containing arsenic or vessels that have contained arsenic compounds are acid cleaned, the very dangerous gas arsine, AsH_3, an analog of ammonia, may be evolved. The gas is colorless with an unpleasant odor resembling that of garlic. Even a small amount of arsine, when inhaled, causes headache, vomiting, abdominal pains, and chills. The inhalation of any significant amount leads to renal failure, pulmonary edema, and death. If there is the slightest possibility of evolving this gas during chemical cleaning, permanganate scrubbers are indicated and self-contained breathing apparatus must be worn at all times. Workers should also be aware that small amounts of arsenic are present as an impurity in many metals, including iron, and it is a common contaminant of commercial mineral acids. Chemical cleaning is normally conducted in an open system unless toxic gas is expected to be evolved. Vented gases are scrubbed with a suitable solution, or else flared, so that the toxic gases hydrogen sulfide, hydrogen cyanide, cyanogen, and arsine are not released to the atmosphere. As pockets of toxic gas can remain trapped in equipment, even after chemical cleaning is completed, self-contained breathing apparatus and protective clothing should be worn when removing blinds and disconnecting hoses and piping preparatory to returning a unit to service.

Acid cleaning of steel always generates some hydrogen, which is vented to the atmosphere occasionally to prevent pressurizing the vessel being cleaned. When cleaning a boiler, for instance, hydrogen gas is evolved so it is important that there be no source of ignition in the immediate work area, particularly an enclosed space such as a boiler house or power plant. Welding, open fires, smoking, or any other source of ignition should not be permitted within 50 ft of a work area in which a flammable solvent is being used. Also, the chemical cleaning unit should be bonded to the equipment being cleaned to avoid static electrical discharge when using solvents other than water-based solutions for cleaning.

Finally, it is essential that workers do not center their attention on one hazard only. If a hazardous gas is being evolved during an acid cleaning, it is important to remain aware of the acid hazard as well as that of the toxic gas. Indeed, to work safely all concerned must remain aware of all hazards connected with a chemical cleaning procedure.

I.4 EQUIPMENT USED BY CONTRACTORS

A chemical cleaning unit consists of a minimum of three tanks mounted on a flat-bed truck; the circulating tank is lined with Hypalon, a synthetic, rubber-like, polymer; the other two are unlined carbon steel. An unlined tank must be used to circulate aromatic solvents, which would destroy the linings of the other tanks. The three tanks are connected in series to provide a supply for circulation, a knock-out or surge tank, and a caustic scrubber, if needed, to absorb acidic gases. The circulating and the surge tanks should be fitted with sight glasses and the system should be so arranged as to allow flushing or draining to a vacuum truck without disconnecting the supply or return lines. When a secondary scrubber is used, the vent line from the primary to the secondary scrubber is equipped with a sampling valve to permit determining when the primary scrubber is saturated, allowing toxic gases to enter the secondary scrubber.

With this equipment a chemical solution can be circulated from the supply tank, through a truck-mounted centrifugal pump and, by hose or temporary piping, into the lowest connection on the vessel to be cleaned, through the vessel and out of the highest connection, then back to the supply tank on the truck. Temporary 2-in. piping should be run from the connections on the unit being cleaned to the vicinity of the truck, then connected to it with hoses that have been tested with water at the maximum pressure of the circulating pump (50–100 psi). A typical arrangement for cleaning multiple heat exchangers is shown in Fig. I.3.

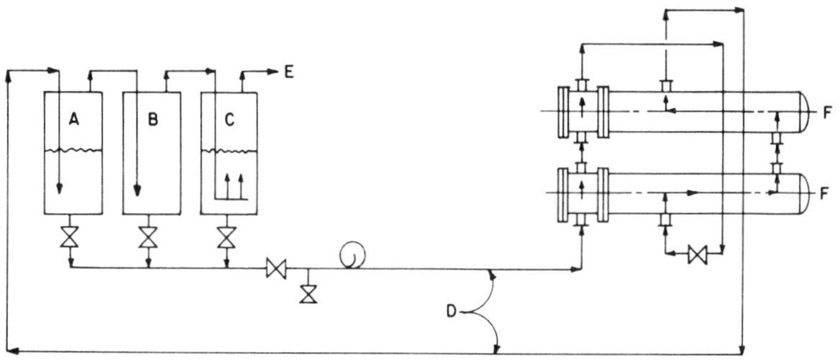

Fig. I.3 Set up for cleaning heat exchangers in series. A – Acid Tank; B – Surge Tank; C – Casutic Scrubber; D – Temporary Piping; E – Vent to Flare; F – Heat Exchangers

Before commencing the cleaning, tanks, hatchcover seals, lines, hoses, valves, and associated equipment should be soap-tested for leaks at 10 psi maximum using nitrogen, carbon dioxide, or air. The entire job area, including the truck, hoses, piping, and the unit to be cleaned, is roped off by the contractor, and approved warning signs must also be posted reading: "CAUTION – CHEMICAL CLEANING – KEEP OUT."

References

(1) Latimer, W. M. 1952. *The oxidation states of the elements and their potentials in aqueous solutions.* 2nd ed. Englewood Cliffs: Prentice-Hall.

(2) Garrels, R. M. and Christ, C. L. 1965. *Solutions, minerals, and equilibria.* New York: Harper and Row.

(3) Van Olphen, H. 1963. *An introduction to clay colloid chemistry*. New York: Interscience.

(4) McCoy, J. W. 1980. *Microbiology of cooling water*. New York: Chemical Publishing Co.

(5) Badger, G. M. and Cook, J. W. 1956. Aromatic compounds with condensed nuclei: fluorene, acenaphthene and polycyclic compounds. Chapter XXII in *Chemistry of the carbon compounds*, Vol. IIIB. E. H. Rodd, ed. Amsterdam: Elsevier Publishing Co.

(5a) Rusin, A. and Kulezycka, J. (Dep. Carbochem., Cent. Min. Inst., Katowice, Pol.) 1975. Effectiveness of the separation of catalytically hydrocracked coal extracts on alumina and Sephadex LH-20. *Chem. Anal.* (Warsaw), 20 (6), 1177 (Pol.) (C.A. 84: 124248r.)

(6) Jones, M. C. K. and Hardy, R. L. 1952. Petroleum ash components and their effect on refractories. *Ind. Eng. Chem.* 44: 2615.

(7) West, J. M. 1971. *Electrodeposition and corrosion processes*. London: Van Nostrand Reinhold Co. Ltd.

(8) Uhlig, H. H. 1967. *Corrosion and corrosion control*. New York: John Wiley & Sons.

(9) Am. Soc. Testing Materials. 1978. Standard reference method for making potentiostatic and potentiodynamic anodic polarization measurements. *Annual Book of ASTM Standards*. Part 10, G-5.

(10) Evans, U. R. 1929. The distribution and the velocity of the corrosion of metals. *J. Franklin Inst.* 208:45.

(11) Tafel, J. 1904. Über die polarisation bei kathodischer wasserstoffentwicklung. *Z. physik. chem.* 50:641.

(12) Stern, M. 1957. Fundamentals of electrode processes in corrosion. *Corrosion* 13:775t.

(13) Am. Soc. Testing Materials. 1977. Corrosivity of solvent systems for removing water-formed deposits. *Annual Book of ASTM Standards*. Part 31, D3263.

(14) Bregman, J. I. and Newman, T. R. 1959. Developments in cooling tower system treatments. *Corrosion* 15:97t.

(15) Freedman, A. J., Troscinski, E. S., and Dravnieks, A. 1958. An electrical resistance method of corrosion monitoring in refinery equipment. *Corrosion* 14:175t.

(16) Marsh, G. A. and Schaschl, E. 1964. Instantaneous corrosion-rate measurements and their use in the study of corrosion inhibitors. *Proc. Am. Petrol. Inst., Sect. III* 44:166.

(17) Am. Soc. Testing Materials. 1977. Tentative methods of test for corrosivity of water in the absence of heat transfer (electrical methods). *Annual ASTM Standards.* Part 31, D2776.

II

Chemicals Used in Industrial Cleaning

A great many different chemicals have been suggested and used for chemical cleaning, including alkalies, acids, inhibitors, wetting agents, sequestrants, and organic solvents. The choice of solvent and auxiliary chemicals is determined by the metallurgical composition of the unit to be cleaned, the solubility of the fouling material, and the cost of the chemicals. Rabald[1] has compiled an exceptionally useful and comprehensive list of chemicals and their effects on metals and alloys, bricks, tiles, cements, rubber, woods, stoneware, and many other materials. Solvents should ideally be selected on the basis of results of solubility tests on samples of the fouling materials, if available; the importance of representative samples of these materials cannot be exaggerated.

II.1 ALKALINE SOLUTIONS

Alkalies and salts of weak acids are used to emulsify, or saponify, oily organic materials, to neutralize acidic residues from chemical cleaning, and to aid in passivating cleaned metal surfaces. These solutions should not be used in equipment containing aluminum, lead, or zinc parts, as above pH 10 these metals are rapidly attacked by hydroxyl ion with the evolution of dangerous amounts of hydrogen, which may be ignited by the heat of the reactions.

$$Al + OH^- + H_2O = AlO_2^- + 3/2H_2 \quad \text{(II-1)}$$

$$Zn + 2OH^- = ZnO_2^- + H_2 \quad \text{(II-2)}$$

$$Pb + OH^- + H_2O = HPbO_2^- + H_2 \quad \text{(II-3)}$$

Similarly, aqueous ammonia must not be used in the presence of copper or its alloys, as ammonium hydroxide, NH_4OH, causes rapid corrosion and stress corrosion cracking of these materials. Also, in aerated solutions, copper is readily oxidized to the cupric ammonia complex.

$$Cu + 4NH_3 + \tfrac{1}{2}O_2 + H_2O = Cu(NH_3)_4^{++} + 2OH^- \qquad \text{(II-4)}$$

II.1.a Inorganic Alkaline Solutions

For convenience and considerations of safety, it is preferable to derive hydroxyl ion for deoiling from the hydrolysis of the salts of weak acids, most commonly, carbonic, phosphoric, and metasilicic acids. Using phosphoric acid as an example, the pH of solutions of trisodium phosphate can be calculated by the following method. Phosphoric acid is a tribasic acid that dissociates in three steps. Parentheses indicate concentrations in mols/liter.

$$H_3PO_4 = H_2PO_4^- + H^+ \qquad K_1 = \frac{(H_2PO_4^-)(H^+)}{(H_3PO_4)} = 7.6 \times 10^{-3} \qquad \text{(II-5)}$$

$$H_2PO_4^- = HPO_4^{--} + H^+ \qquad K_2 = \frac{(HPO_4^{--})(H^+)}{(H_2PO_4^-)} = 6.3 \times 10^{-8} \qquad \text{(II-6)}$$

$$HPO_4^{--} = PO_4^{-3} + H^+ \qquad K_3 = \frac{(PO_4^{-3})(H^+)}{(HPO_4^{--})} = 4.4 \times 10^{-13} \qquad \text{(II-7)}$$

As HPO_4^{--} is a very weak acid, it tends to be formed by the hydrolysis of PO_4^{-3} when the latter, in the form of trisodium phosphate, is dissolved in water. This reaction also simultaneously releases a hydroxyl ion from the water.

$$PO_4^{-3} + H_2O = HPO_4^{--} + OH^- \qquad \text{(II-8)}$$

The hydrolysis constant can be calculated from K_3 and the ion product of water, $K_w = (H^+)(OH^-) = 10^{-14}$.

$$\frac{(HPO_4^{--})(OH^-)}{(PO_4^{-3})} = K_h = K_w/K_3 = 2.3 \times 10^{-2} \qquad \text{(II-9)}$$

The concentration of hydroxyl ion in a solution of trisodium phosphate can be calculated by noting that it is equal to that of HPO_4^{--} and that the concentration of PO_4^{-3} is diminished by the concentration of OH^- produced, as is apparent from Eq. (II-8). Therefore,

$$\frac{(OH^-)^2}{(PO_4^{-3}) - (OH^-)} = 2.3 \times 10^{-2} \qquad \text{(II-10)}$$

and,

$$(OH^-)^2 + 2.3 \times 10^{-2}(OH^-) - 2.3 \times 10^{-2}(PO_4^{-3}) = 0 \qquad \text{(II-11)}$$

Suppose that a 1 percent solution of trisodium phosphate is prepared, and it is desired to know its pH.

$$1\%\ Na_3PO_4 = 10\text{ g/l} = (10/164)\text{ mols/liter} = 6.1 \times 10^{-2}$$

Trisodium phosphate solutions are usually prepared from dodecahydrate, $Na_3PO_4 \cdot 12H_2O$, the most common commercial form. If this salt is used 23 g of it are required to make 1 liter of 1% Na_3PO_4, i.e., 6.1×10^{-2} (mol/l) $\times$ 380 (g/mol). Substituting in Eq. (II-2) and applying the quadratic formula,

$$(OH^-) = \frac{-2.3 \times 10^{-2} + \sqrt{(2.3 \times 10^{-2})^2 + 4 \times 2.3 \times 10^{-2} \times 6.1 \times 10^{-2}}}{2}$$

$$= 2.79 \times 10^{-2}$$

$$pOH = -\log_{10}(2.79 \times 10^{-2})$$

$$= 1.55$$

$$pH = 14 - pOH$$

$$= 12.4$$

Thus, a 1 percent solution of trisodium phosphate is about 0.03*N* in hydroxyl ion as the result of hydrolysis. That is, Eq. (II-8) goes to the right, as written, until the concentration of OH^- reaches about 0.03*N*. If, in use, the concentration of hydroxyl ion decreases, more phosphate ion hydrolyzes until the concentration of hydroxyl ion is again 0.03*N*. Sodium phosphate thus serves as a reservoir of hydroxyl ion, which is the active ingredient in dispersing and emulsifying oily residues. The advantage of this is that the solution never contains a concentration of hydroxyl ion higher than 0.03*N* (equivalent to 0.12 percent NaOH), nor does the pH exceed 12.4, as would be the case if sodium hydroxide were used alone.

BOIL-OUT SOLUTIONS

These are alkaline solutions used to remove various organic contaminants from new equipment, especially drum boilers or retubed heat exchangers, that is to be chemically cleaned. In addition to mill scale, new boilers and other vessels may contain cutting oils, grease, lubricating oils, pipe threading compounds, drawing compounds, slushing compounds, and various protective coatings. To remove these materials, new drum boilers are subjected to an alkaline boil-out to expose the surface of the metal so that a normal uniform protective film of magnetite can form when the unit is put into service. If preoperational cleaning is neglected, pitting and generalized corrosion occur because covered areas of metal are anodic to clean areas.

For many years a favorite combination for conditioning new drum boilers was 3000 ppm of NaOH and 3000 ppm of Na_2CO_3. The mixed solvent was pumped into the boiler, which was then fired for 24–48 h at about 100 psi, during which time the concentration of the solution was occasionally checked and strengthened, if necessary. The addition of sodium nitrate has sometimes been recommended to protect steel from caustic embrittlement. Procedures are given in Chapter XI for checking the concentrations of NaOH and Na_2CO_3 in boil-out solutions.

Hantelman[2] recommends a mixture of Na_3PO_4 and Na_2HPO_4 for the preoperational deoiling of preboiler systems and boilers that he found especially effective for removing tenacious protective coatings. A solution containing 7000 ppm Na_3PO_4, 3000 ppm Na_2HPO_4, and 0.2 percent of wetting agent is heated to 180°F for 12 h. Should the concentration of phosphate fall below 3000 ppm PO_4^{-3}, circulation must be continued for 48 h. Atwood,[30] for the same application, suggests Na_3PO_4, NaOH, and Na_2CO_3 in some optional combination at 0.5-1.0 percent total concentration, with the addition of 0.1–0.25 percent of a nonionic or cationic wetting agent. This alkaline flush is pumped through the condensate and feed water systems, then back through the shells of the feed water preheaters to the pump suction at the condenser hot well. A similar solution is used for boiler boil-out, or alkaline wash in once-through boilers. Duncan, et al.[4] specify an alkaline flush of preboiler systems. Their solution contains 800 ppm PO_4^{-3} and 2500 ppm OH^-; it is circulated at 175°F for 4 h. Hoppe and Parmley[5] prefer to avoid NaOH and Na_2CO_3; they favor Na_3PO_4 or a combination of Na_3PO_4 and Na_2HPO_4, with a wetting agent added, for flushing once-through boilers.

In addition to the four alkaline salts already mentioned, sodium metasilicate is also often used for deoiling contaminated equipment, either alone or in combination with one of the phosphates, or with sodium carbonate. As all of the salts thus far mentioned function as deoiling agents by releasing hydroxyl ion by hydrolysis, it seems reasonable to infer that one of them could be selected as the optimum performer. Table II-1 displays data pertinent to making this selection.

TABLE II-1

Properties of Alkaline Chemicals

Chemical	*pH of 1 percent solution*	*Concentration of* OH^-, *N*
Sodium carbonate	11.7	0.005
Trisodium phosphate	12.4	0.03
Sodium metasilicate	12.9	0.08
Sodium hydroxide	13.4	0.25

The pH of both sodium metasilicate and sodium hydroxide, at 1 percent, is greater than 12.5, above which value a used solution is classified as a hazardous waste, requiring special disposal. Furthermore, many boiler operators prefer to avoid using sodium hydroxide, particularly for washing preboiler systems, because of the possibility of stress corrosion cracking in the highly alkaline solution. Sodium carbonate at a concentration of 1 percent is only 0.005*N* in hydroxyl ion, whereas trisodium phosphate at the same concentration has a pH of 12.4, with a reasonably high normality in hydroxyl ion of 0.03*N*. In addition, phosphate ion, in the presence of oxygen, has definite value as a passivating agent, a subject considered at length in Chapter V. It should be further noted that a 10 percent solution of sodium carbonate has a pH of 12.0, and is about 0.01*N* in hydroxyl ion. That is, a ten-fold concentration of Na_2CO_3 produces only one-third of the concentration of hydroxyl ion in a 1 percent solution of Na_3PO_4. Thus, it would appear that a 1 percent solution of trisodium phosphate, with perhaps the addition of 0.1–0.2 percent of a nonionic wetting agent, should be the solution of choice for routine deoiling. Four to six hours at 180°F usually suffices if the oil is at all extractable.

ALKALINE PERMANGANATE

When polymeric or tarry materials are encountered that do not yield to the solutions just described, it may be necessary to resort to alkaline permanganate. An examination of the half-reactions for the oxidation of iron by permanganate in alkaline solution suggests that this oxidizing agent might be corrosive to steel.

$$Fe + 3OH^- = Fe(OH)_3 + 3e^- \qquad E_B^\circ = 0.771\ V \qquad (II\text{-}12)$$

$$MnO_4^- + 2H_2O + 3e^- = MnO_2 + 4OH^- \qquad E_B^\circ = 0.588\ V \qquad (II\text{-}13)$$

The complete reaction is

$$MnO_4^- + 2H_2O + Fe = MnO_2 + Fe(OH)_3 + OH^-$$

$$E_B^\circ = 1.359\ V \qquad (II\text{-}14)$$

Because of the passivating effect of the strong oxidizing agent, however, corrosion rates, as determined in the laboratory, are about 9 mpy (0.023 lb/ft^2-day) on carbon steel, and 20 mpy (0.052 lb/ft^2-day) on admiralty brass. These rates are, of course, negligible in a chemical cleaning procedure, which would rarely exceed 24 hours' duration.

In practice, a solution containing 1 percent NaOH and 0.7 percent $KMnO_4$ is pumped into the contaminated vessel and heated at 150–175°F for 12–24 h. When cleaning a boiler, it may be very lightly fired (flue gas temperature 250°F, maximum), or heated with igniters or warm-up guns, with the steam drum vented. At the conclusion of the treatment the boiler is cooled, drained, and flushed with deaerated water, after which it may be inspected and acid-cleaned, if necessary.

Carbonized or coke deposits are best removed by chromic acid, as described in Section II.2. Asphaltic and other tarry fouling ordinarily yields readily to certain nonaqueous solvents discussed in Chapter III.

INDUSTRIAL CLEANERS AND DETERGENTS

Hundreds of proprietary cleaners are commercially available, formulated for specific applications. The type used for so-called heavy duty industrial cleaning may contain mixtures of the inorganic salts already mentioned, with the addition of surface-tension depressants, wetting agents, penetrants, and emulsifiers. These solutions are usually meant to be diluted, heated with steam, and applied for general cleanup through a high-pressure hose fitted with a special nozzle containing a mixing jet that automaticaly regulates the flows of steam, water, and detergent. Additives include sodium sulfonates, alkyl aryl sulfonates, alkylene glycol alkyl ethers, sodium dioctylsulfosuccinate, and dozens of others.

It is advisable to perform some screening tests on samples of industrial cleaners before contracting for large quantities of them. The pH of a 5 percent solution should be less than 12, and the wetting time, described in Chapter XI, should be not more than 30–40 s. If these preliminary results prove satisfactory, the product should be field-tested at various concentrations ranging from 2.5 to 10 percent to provide a basis for comparing its cost and effectiveness with other products. In general, these commercial cleaners are not circulated through industrial process units; they are intended primarily for cleaning the exterior of vessels, the walls of tanks, cement floors, and the like.

II.1.b Alkaline Emulsions

In some instances, iron oxides, sulfides, and other inorganic deposits are coated with oily material that shields them from the solubilizing action of acidic cleaning solutions. If the oil is light-bodied, e.g., a lubricating or turbine oil, the trisodium phosphate-wetting agent formulation previously mentioned can be circulated for a few hours at 200°F. If, however, the organic material is heavy, waxy, greasy, polymeric, or present in relatively large amounts, a hydrocarbon such as automotive diesel fuel, kerosene, or an aromatic solvent can be added to the phosphate solution to form an emulsion that disperses and emulsifies the organic fouling. Hydrocarbon solvents are added in the proportion of 20–80 percent by volume of the total volume of cleaning solution, together with 0.1 percent of an oil-soluble, nonionic surfactant to stabilize the emulsion. A suitable emulsifying agent is nonylphenoxypolyoxyethylene ethanol,* which is discussed further in Section II.4. If more than 0.1 percent of the emulsifier is used in the alkaline emulsion it is likely that the hydrocarbon will not separate from the aqueous phase in a reasonable length of time. The emulsion is circulated at 140°F for 4–6 h through the equipment to be deoiled, after which the mixture is transferred to a vacuum tank truck, where the two layers are allowed to separate. The aqueous phase is then drained out of the tank truck and the remaining hydrocarbon layer is transferred to a

* Manufactured and sold under the tradename Igepal CO-430 by GAF Corporation, 140 West 51st Street, New York, N.Y. 10020.

recovered oil tank for disposal or reprocessing. If a small amount of water-soluble surfactant, say 10 ppm, is also included in the alkaline emulsion, subsequent rinsing of the equipment is often facilitated.

II.2 ACIDIC SOLUTIONS

Except for a few amphoteric or acidic substances such as $Al(OH)_3$, ZnO, and a few alkaline earth silicates, very little of the inorganic portion of a fouling deposit is solubilized by the alkaline treatments discussed in Section II.1, which, of course, are applied primarily to remove oil and other organic material. Just as alkalies tend to dissolve acidic material, so do acids attack basic material and they are, therefore, appropriate for dissolving many inorganic oxides and salts of weak acids. As pointed out in Chapter I, all acids are inherently hazardous to handle, especially when hot, so the safety of workmen and equipment must be the primary concern in acid cleaning. Other important considerations are minimizing the creation of hazardous wastes, which entail a great deal of expense in their disposal, and ensuring that the cleaning procedure selected will be quick and effective, so that a satisfactory start-up or return to service of the equipment to be cleaned can be anticipated.

If possible, the acidic solvent is chosen on the basis of solubility tests on the fouling material to be removed, with due attention also being given to metallurgical considerations. Although preliminary deoiling exposes inorganic material to the action of acids, it must be realized that blocked passages cannot be opened by chemical solutions, because they cannot flow through. In such cases resort must be made to one or more of the mechanical methods discussed briefly in Chapter IV.

Acid cleaning inevitably leaves a large mass of insoluble particulate matter that is most easily removed by high-pressure water jets. This, in general, is done immediately after draining and neutralizing any acid remaining in the equipment. As procedures vary, depending upon the unit being cleaned, typical examples are given in Chapters VII–X; here a number of individual acids, their chemical properties, and their applications are described.

II.2.a Inorganic Acids

The inorganic acids, often called mineral acids, that are most useful in chemical cleaning are hydrochloric, sulfuric, phosphoric, sulfamic, and occasionally, hydrofluoric acids. With the exception of the last, all of these are strong acids, in the sense that they are highly dissociated, yielding a high concentration of hydrogen ion per mole. Thus, hydrochloric acid in dilute solutions is almost completely dissociated into hydrogen ion and chloride ion.

$$HCl\ (aq) = H^+ + Cl^- \tag{II-15}$$

That is to say, when hydrogen chloride gas dissolves in water it is essentially completely dissociated into ions, as shown in Eq. (II-15), with virtually no undissociated HCl molecules in the solution. In contrast, when hydrogen fluoride dissolves in water it is only partially dissociated into hydrogen and fluoride ions, the extent being quantitatively expressed by its acid dissociation constant.

$$K_a = \frac{(H^+)(F^-)}{(HF)} = 6.6 \times 10^{-4} \tag{II-16}$$

The concentration of H^+ in a solution of this moderately weak acid can be calculated by noting that

$$\frac{(H^+)^2}{(HF) - (H^+)} = 6.6 \times 10^{-4}$$

In a 1 percent solution of HF, for example, the concentration of H^+ is calculated as follows:

$$(HF) = 10(g/l)/20(g/mol) = 0.5\ mol/l$$

$$(H^+)^2 + 6.6 \times 10^{-4}(H^+) - 0.5 \times 6.6 \times 10^{-4} = 0$$

$$(H^+) = \frac{-6.6 \times 10^{-4} + \sqrt{(6.6 \times 10^{-4})^2 + 4 \times 0.5 \times 6.6 \times 10^{-4}}}{2}$$

$$= -6.6 \times 10^{-4} + 3.6 \times 10^{-2}$$

$$= 1.8 \times 10^{-2}$$

Hydrofluoric acid at a concentration of 1 percent is, thus, $1.8 \times 10^{-2} \times 100/0.5$, or 3.6 percent dissociated; the pH of the solution is 1.7.

Most inorganic acids are used when there is a need for high concentrations of hydrogen ion, whereas the weak organic acids are used in applications in which a moderate, but continuously replaceable, supply of hydrogen ions is desired. Hydrofluoric acid is valuable for its specific capability of dissolving silicates.

HYDROCHLORIC ACID

The majority of corrosion fouling and water-formed deposits yield to hot, dilute hydrochloric acid, for which reason it is the most frequently used acidic solvent in chemical cleaning. All acids must be suitably inhibited before being introduced into industrial equipment to minimize acidic attack on the equipment itself. The aim, of course, is to dissolve or disperse as much fouling as possible without excessively corroding the vessel that is fouled. Inhibitors and auxiliary additives are discussed in Sections II.3-II.5.

Hydrochloric acid, at concentrations of 5-10 percent by weight, is effective for dissolving, either partially or completely, ferric oxide, Fe_2O_3, magnetite (mill scale), Fe_3O_4, ferric phosphate, $FePO_4$, ferrous sulfide, FeS, calcium carbonate, $CaCO_3$, hydroxyapatite, $Ca_{10}(PO_4)_6(OH)_2$, and magnesium phosphate, $Mg_3(PO_4)_2$. In addition, it disintegrates masses of algae and microbiological slime sometimes found in water-cooled heat exchangers and in compressor water jackets. The acid is ineffective against anhydrite, $CaSO_4$, silicates (except sodalite, $Na_2O \cdot 3Al_2O_3 \cdot 6SiO_2 \cdot NaCl$), clays, sand, wood chips, coke, fused sulfur, metallic copper, and pyrite, FeS_2. Hydrochloric acid is the least expensive of the acids used for cleaning, and is especially suited for dissolving ferric oxide because the driving force of the neutralizing reaction

$$Fe_2O_3 + 6H^+ = 2Fe^{+3} + 3H_2O \qquad \text{(II-17)}$$

obtained with any acid, is augmented by the formation of the stable hexachloroferric ion.

$$Fe_2O_3 + 6H^+ + 12Cl^- = 2FeCl_6^{-3} + 3H_2O \qquad \text{(II-18)}$$

The effects of both hydrogen and chloride ions are thus brought into play in dissolving this oxide in HCl. Although the hexachloroferric ion is quite stable, it is also notably labile, in that it rapidly exchanges with radioactive chloride in aqueous solution.

When properly inhibited, hydrochloric acid is appropriate for cleaning carbon steel, low-chromium steel, cast iron, admiralty brass, bronze, cupronickel, and monel. Corrosion rates during cleaning depend upon the metal being cleaned, the concentration of inhibitor, the concentration of the acid, the time of exposure, and the temperature and velocity of the acidic solution. As the corrosivity of HCl increases very rapidly with velocity, this should be limited to 1-2 fps. Atwood[3] found the cleaning capacity of HCl to be quite satisfactory when applied by the fill-and-soak method, that is, with no circulation. Klein and Atwood[6] earlier reported that 5 percent HCl heated to 160–170°F and circulated at 2 fps, removed 5–10 mils of magnetite within 30 min. They also found that none of the acidic solvents discussed in this section was capable of removing mill scale in the presence of petroleum-based coatings.

The corrosion rate of carbon steel in uninhibited 10 percent hydrochloric acid at 160°F is about 16–20 mpd; with 0.1–0.5 percent of a suitable inhibitor it is 0.7–0.8 mpd in 5–15 percent HCl; in 10 percent HCl at temperatures ranging from 140 to 200°F the rate is 0.5–1.0 mpd. In the presence of ferric ion at a concentration of 0.5 percent, the corrosion rate of iron in inhibited 10 percent HCl increases five- to ten-fold. Ferric ion rapidly and uniformly etches both copper and ferrous alloys on account of the very high change in free energy accompanying the reaction

$$2Fe^{+3} + Fe = 3Fe^{++} \tag{II-19}$$

the equilibrium constant of which is $K = 1 \times 10^{41}$. This shows that scarcely any ferric ion can exist in equilibrium with metallic iron.

Hydrochloric acid should not be used to clean stainless steels, titanium, zinc, aluminum, or galvanized iron. Engle, et al.[7] have shown that austenitic stainless steels are not attacked by inhibited hydrochloric acid at 150°F. After neutralization, however, chloride ion at concentrations as low as 40 ppm can cause intergranular or transgranular cracking, as well as steep-walled pitting in stressed areas. The chloride ion, in neutral solution, is strongly adsorbed on the surface of ferrous metals, where it destroys passivity by displacing oxygen, penetrating defects in the normally protective film, and lowering the hydrogen overvoltage; cracking then ensues at areas stressed well below the yield point. For this reason, hydrochloric acid is avoided in chemically cleaning stainless steels.

Although titanium is resistant to salt water erosion, hydrogen sulfide, ammonia, nitric acid, chloride, and chlorine, it is susceptible to hydrogen embrittlement, and corrodes rapidly in boiling hydrochloric acid.[8] Aluminum and zinc, of course, are rapidly dissolved by any mineral acid. Some hydrogen compressors contain thin, precisely machined aluminum labyrinth rings that can tolerate no metal loss. Acids must never be used to clean these machines; they can be safely cleaned with a neutral detergent solution.

When cleaning very large equipment, e.g., a drum boiler, which might require 25,000 gal or more of dilute acid, it is necessary to blend-fill. In this procedure the required amount of inhibited concentrated acid and dilution water are separately and simultaneously metered into the empty boiler at such a rate that not more than two hours elapse during the filling. For smaller vessels, such as heat exchangers or surface condensers, the dilution is made in the recirculating tank on the chemical cleaning truck. In practice, the proper amount of inhibited concentrated acid is added to the tank, then dilution water is added at the job site through a mixing manifold.

Commercial grades of hydrochloric acid (frequently called muriatic acid) are available at concentrations of 28, 32, and 37 weight percent of HCl, which are diluted with water to use concentrations of 5, 7.5, or 10 percent. The proper volumes of concentrated acid and water to use can be calculated approximately from the specific gravities of the solutions and the percentage of HCl in each. There is a small contraction in volume on mixing, but this is of no consequence as the strengths of the solutions are always checked by titration by the method in Chapter XI, while they are being prepared. Specific gravities and percentages can be obtained by consulting a chemistry handbook.(9) The volume of concentrated acid required to make 1000 gal of dilute acid is calculated by the following formula.

$$V_{(\text{conc.})} = \frac{\text{sp. gr. (dilute)} \times \%\,\text{(dilute)} \times 1000\ \text{(gal)}}{\text{sp. gr. (conc.)} \times \%\,\text{(conc.)}}$$

Thus, if it is desired to prepare 600 gal of 5.0 percent HCl, specific gravity 1.023 from 32 percent HCl, specific gravity 1.159, the volume of concentrated acid to be diluted to 600 gal is

$$\frac{1.023 \times 0.050 \times 600}{1.159 \times 0.32} = 82.7\ \text{gal}$$

The values in Table II-2 were calculated in this way.

To correct the concentration of a prepared acidic cleaning solution it is more convenient to use normality (equivalents/liter) than percentage (pounds/hundred pounds). Normality is determined by measuring a sample of dilute acid with a transfer pipet, then titrating it with a standard solution of sodium hydroxide to the end point of an appropriate indicator, as described in Chapter XI. Table II-3 lists normalities corresponding to various percentages of hydrochloric acid in the ranges normally used for chemical cleaning; other values are given in chemical handbooks.(8)

TABLE II-2

Preparation of Hydrochloric Acid Solutions

Concentrated HCl		*Gallons of concentrated to prepare* 1000 gal *of dilute*					
		5% HCl		7.5% HCl		10% HCl	
% HCl	Sp. gr. @ (20/4)°	Vol. (conc.)	Sp. gr. (dilute)	Vol. (conc.)	Sp. gr. (dilute)	Vol. (conc.)	Sp. gr. (dilute)
28	1.139	160	1.023	243	1.035	328	1.047
32	1.159	138	1.023	209	1.035	282	1.047
37	1.184	117	1.023	177	1.035	239	1.047

TABLE II-3

Strengths of Hydrochloric Acid Solutions

% HCl, *by weight*	*Normality, equivalents/liter*	*Sp. gr.* @ (20/4)°C
4	1.12	1.018
5	1.40	1.023
6	1.69	1.028
7	1.99	1.033
8	2.28	1.038
9	2.58	1.040
10	2.87	1.047
28	8.75	1.139
32	10.2	1.159
37	12.0	1.184

In correcting the strength of a prepared dilute solution of any acid two cases arise:

1. *Normality is greater than desired.*

In this case dilute the prepared solution according to the following formula:

$$V_w = V_s\left(\frac{N_s - N_d}{N_d}\right) \tag{II-20}$$

Where:

V_w = volume of water to be added, gallons
V_s = volume of solution that is too strong, gallons
N_s = normality of strong solution, equivalents/liter
N_d = desired normality (from Table II-3), equivalents/liter

2. *Normality is less than desired.*

In this case the strength of the weak prepared solution is increased by adding concentrated acid according to the following formula:

$$V_s = V_w \left(\frac{N_d - N_w}{N_s - N_d} \right) \tag{II-21}$$

Where:

V_s = volume of concentrated acid to be added, gallons
V_w = volume of weak prepared solution to be strengthened, gallons
N_d = normality desired (from Table II-3), equivalents/liter
N_w = normality of weak solution, equivalents/liter
N_s = normality of concentrated acid, equivalents/liter

Suppose, for example, that an attempt is made to prepare 500 gal of a 5.0 percent solution of hydrochloric acid, but it is found by titration that its normality is 1.60. The amount of water to be added to reduce the normality to 1.40, equivalent to 5.0 percent by weight, is

$$V_w = 500 \left(\frac{1.60 - 1.40}{1.40} \right) = 500 \times 0.143 = 71.4 \text{ gal}$$

Similarly, if the normality were found to be 1.18, then the volume of 32 percent HCl to add is

$$V_s = 500 \left(\frac{1.40 - 1.18}{10.2 - 1.40} \right) = 500 \times 0.025 = 12.5 \text{ gal}$$

As these correction formulas apply to all solutions of acids or alkalies, they are not repeated in the following pages.

If it is necessary to neutralize spent hydrochloric acid solutions before they are discarded, a variety of bases are available for this purpose. Limestone, for instance, reacts rapidly with hydrochloric acid according to Eq. (II-22).

$$2HCl + CaCO_3 = CaCl_2 + H_2O + CO_2 \tag{II-22}$$

The percentage of HCl in a spent solution is determined by titration and the weight of acid in the solution, in pounds per gallon, is obtained from a chemical handbook.[9] The number of pounds to be neutralized is calculated by multiplying the latter number by the number of gallons to be neutralized. The number of pounds of limestone (or other base) required to neutralize that number of gallons is then calculated using the molecular weight relations in Eq. (II-22).

Suppose, for example, that 1000 gal of a 2 percent solution of hydrochloric acid is to be neutralized with limestone. From a handbook it is found that 2 percent HCl contains 0.168 lb of HCl per gallon of solution. A thousand gallons, therefore, contains 1000 × 0.168, or 168 lb. From Eq. (II-22), 73 lb of HCl reacts with 100 lb of $CaCO_3$. Therefore, 168 lb of HCl requires 168 × 100/73, or 230 lb of limestone to neutralize it. Other values in Table II-4 are calculated in the same way.

TABLE II-4

Neutralization of Spent HCl Solutions

		Pounds of base to neutralize 1000 gal				
% HCl	lb HCl/ 1000 gal	Limestone ($CaCO_3$)	Hydrated lime ($Ca(OH)_2$)	Soda ash (Na_2CO_3)	Caustic soda (NaOH)	Gal 50 Bé NaOH
1	84	115	85	122	92	14.5
2	168	230	170	244	184	28.9
3	254	348	257	369	278	43.7
4	340	466	345	494	373	58.6
5	427	585	433	620	468	73.5

As the vapor pressure of hydrogen chloride is high in a concentrated solution of hydrochloric acid, it fumes in moist air and is highly corrosive to the respiratory tract. For precautions in handling this chemical see Section I.3.

SULFURIC ACID

Because of its high heat of dilution and its devastating effect on animal tissues, sulfuric acid is avoided by most chemical cleaning contractors. As it is used occasionally, however, it deserves a brief discussion. The acid, which is also called oil of vitriol, is available in three commercial strengths: 60.0 Bé, 77.8 percent; 66.0 Bé, 93.2 percent; and 66.4 Bé, 98.0 percent. As the acid becomes very hot when diluted with water, it is advisable to use the lowest strength available, namely, 77.8 percent H_2SO_4. Furthermore, the concentrated acids must *always* be poured slowly into the dilution water rather than the reverse, to avoid spattering and boiling. The acid is used in concentrations ranging from 5 to 15 percent at temperatures up to, but not exceeding, 180°F. It is sometimes used for cleaning stainless steel vessels and for pickling steel piping; for the latter purpose a 10 percent solution is heated to 150°F and applied for 4–6 h. Dilute, inhibited sulfuric acid can also be used on carbon steel, cupro-nickel, admiralty brass, aluminum bronze, and monel. It is not useful for removing aged iron oxides or water-formed deposits, and it rapidly dissolves aluminum and zinc; at concentrations over 3 percent it is corrosive to titanium.

The values in Table II-5 were calculated using the method explained previously for hydrochloric acid solutions.

TABLE II-5

Preparation of Sulfuric Acid Solutions

Concentrated H_2SO_4			*Gallons concentrated to prepare 1000 gal of dilute*					
			5% H_2SO_4		10% H_2SO_4		15% H_2SO_4	
%H_2SO_4	°Be	*(60/60)°F	Vol. (conc.)	Sp. gr. (dil.)	Vol. (conc.)	Sp. gr. (dil.)	Vol. (conc.)	Sp. gr. (dil.)
77.8	60.0	1.705	38.9	1.032	80.3	1.066	124	1.102
93.2	66.0	1.835	30.1	1.032	62.3	1.066	96.6	1.102
98.0	66.4	1.845	28.5	1.032	59.0	1.066	91.4	1.102

* Calculated from the definition of Bé gravity: Sp. gr. @ 60°F = 145/(145 − Bé).

Table II-6 contains values to use for correcting prepared solutions according to the formulas in the preceding section on hydrochloric acid.

TABLE II-6

Strength of Sulfuric Acid Solutions

% H_2SO_4	*N* H_2SO_4	% H_2SO_4	*N* H_2SO_4
1	0.205	10	2.17
2	0.413	11	2.41
3	0.623	12	2.64
4	0.836	13	2.88
5	1.05	14	3.13
6	1.27	15	3.37
7	1.49	77.8	27.0
8	1.72	93.2	34.8
9	1.94	98.0	36.7

Because of the small solubility of $CaSO_4$, neither lime nor limestone is especially suited for neutralizing sulfuric acid solutions; it is best to use soda ash or caustic soda for this purpose. Eq. (II-23) shows that 98 lb of sulfuric acid are neutralized by 106 lb of soda ash.

$$H_2SO_4 + Na_2CO_3 = Na_2SO_4 + H_2O + CO_2 \qquad \text{(II-23)}$$

Suppose, for instance, that 500 gal of 3 percent H_2SO_4 are to be neutralized with soda ash. Reference to a chemistry handbook[9] reveals that 1 gal of 3 percent H_2SO_4 contains 0.255 lb of H_2SO_4, so 500 gal contains 127.5 lb. Therefore, the soda ash needed is

$$127.5 \times 106/98 = 137.9 \text{ lb } Na_2CO_3$$

Every precaution must be taken to protect workers from the hazards of sulfuric acid; it is an extremely dangerous chemical.

CHROMIC ACID

Unlike the other chemicals discussed in this section, chromic acid is used, not as an acid, but as an oxidizing agent. The principal use of chromic acid is for oxidizing binders that hold particles of coke together so that the loosened particles can then be flushed away easily. It is also useful for oxidizing and dissolving pyrite, FeS_2, which is not affected by the mineral acids; as the sulfide is oxidized to sulfate, there is no evolution of hydrogen sulfide in the process.

Cokes are hydrocarbons of high molecular weight, rich in carbon and poor in hydrogen, that are laid down in the course of the decomposition of petroleum, such as occurs in thermal cracking and, inadvertently, in the flash zone of vacuum crude distillation columns. Coking is also encountered in the tubes of furnaces used to preheat petroleum for distillation, but this is removed by steam-air decoking, as described in Chapter I. When sulfuric acid sludge is retorted, a coke is formed that is pyrophoric in air.[(10)]

The disintegration of coke residues by chromic acid is only practicable on a relatively small scale, as, for example, in crude feed preheat exchangers in which a side cut from a crude petroleum distillation column is used to heat feed to the column. Usually, the feed is in the tubes and the side cut is in the shell of these heat exchangers. As the tubes become coked, the pressure drop across the exchanger increases, while the heat transfer decreases, necessitating an increase in the amount of fuel used to raise the temperature of the crude oil. Eventually, when the furnace can no longer make up the heat lost in the preheat exchange, the throughput of crude oil must be reduced. If the tubes are not completely blocked, circulation of a 10–20 percent solution of chromic acid through the tubes for 12–24 h at 190–200°F often disintegrates a coke deposit. Note that the coke is not itself appreciably oxidized by chromic acid, but the polymeric material that binds coke particles together is destroyed. Axsom[(11)] has reported satisfactory results in cleaning this type of heat exchanger with a proprietary solvent, the ingredients of which are unspecified.

Chromic acid must never be used in the presence of easily oxidizable materials, e.g., alcohols, ketones, unsaturates, and the like, as these compounds react violently. Chromic acid is incompatible with virtually all organic material because of its large oxidation potential and, therefore, it cannot be conveniently inhibited. For this reason, it must not be used for cleaning copper, brasses, bronze, aluminum, zinc, or cast iron, for all of these are rapidly attacked.

Chromic acid solutions in concentrations from 10 to 20 percent are safe for carbon steel and stainless steels, but they should not be introduced into equipment before making solubility tests with the actual fouling material that they are expected to remove. If a preliminary sample cannot be obtained, these oxidizing solutions should not be applied. Also, steel piping must be used exclusively for transferring chromic acid solutions, for hoses, other than teflon, are quickly destroyed. After treatment, the solution is drained and the unit is neutralized with an alkaline rinse, then flushed with water. Special care is required in disposing of chromate solutions as they are toxic to marine life.

Chromium trioxide, CrO_3, which is freely soluble in cold water, is the anhydride of dichromic acid, $H_2Cr_2O_7$.

$$2CrO_3 + H_2O = 2H^+ + Cr_2O_7^{--} \quad \text{(II-24)}$$

The anhydride is sold in the form of red crystals from which solutions are prepared, by weight. To prepare a 10 percent solution, for example, proceed as follows (chemical formulas represent pounds):

$$\begin{aligned} CrO_3/(CrO_3 + H_2O) &= 0.10 \\ (CrO_3 - 0.10CrO_3) &= 0.10(H_2O) \\ 0.90(CrO_3) &= 0.10(H_2O) \end{aligned}$$

To prepare 1000 gal of a 10 percent solution:

$$\begin{aligned} 1000 \times 8.33 &= 8330 \text{ lb of water} \\ \text{pounds of } CrO_3 &= 8330 \times 0.10/0.90 \\ &= 925 \end{aligned}$$

Therefore, add 925 lb of CrO_3 to 1000 gal of water, and check the concentration of CrO_3 by the method in Chapter XI.

PHOSPHORIC ACID

This tribasic acid can be used on carbon steel, 300 series austenitic stainless steels, copper, and admiralty brass; it should not be used for cleaning cast iron, 410 stainless steel, aluminum, zinc, or titanium. Many years ago, phosphoric acid was preferred for removing mill scale[12] from new boilers because it leaves the metallic surface more-or-less passive. In addition, a 5 percent solution can be boiled by direct firing without seriously attacking steel, and the higher temperature is a definite advantage. Klein and Atwood[6] found that 3 percent H_3PO_4 heated to 212°F and circulated at 2 fps, required 4 h to remove a 5–10 mil layer of magnetite from a steel surface. Later, Atwood[3] reported that 3 percent H_3PO_4 at 212°F removed a similar layer of magnetite in 6 h without circulation.

Two percent phosphoric acid is sometimes used to pickle and passivate steel piping; it is applied at 120°F for 4–6 h. As the dilute acid has a limited capacity for dissolving oxides of iron, as compared with hydrochloric acid, it is now more often used at higher temperatures to clean austenitic stainless steels, mainly because it is virtually chloride-free, and thus, poses no threat of initiating stress-corrosion cracking in these alloys.

TABLE II-7

Preparation of Phosphoric Acid Solutions

Concentrated H_3PO_4		*Gallons of concentrated to prepare* 1000 gal *of dilute*					
		3%		5%		10%	
% H_3PO_4	Sp. gr. @ (20/4)°F	Vol. (conc.)	Sp. gr. (dil.)	Vol. (conc.)	Sp. gr. (dil.)	Vol. (conc.)	Sp. gr. (dil.)
75	1.579	25.7	1.015	43.3	1.015	88.9	1.053
85	1.689	21.2	1.015	35.7	1.025	73.3	1.053

SULFAMIC ACID

Although more expensive than the other acids thus far mentioned, sulfamic acid, HSO_3NH_2, offers several advantages to those who prefer to do in-house chemical cleaning; that is, without a contractor. It is a white, crystalline solid that, when dry, has negligible effect on the skin or clothing; it is easily handled in the dry state, facilitating weighing and the preparation of solutions of accurate concentrations; it produces no localized etching or pitting. When prepared in combination with sodium chloride (10% HSO_3NH_2, 5% NaCl), the solution is moderately effective for dissolving ferric oxide.

When carefully inhibited, sulfamic acid can be used to clean carbon steel, copper, admiralty brass, and, without chloride, austenitic stainless steels. It is not recommended for aluminum, zinc, cast iron, or 410 stainless steel. The acid hydrolyzes in hot aqueous solution to ammonium bisulfate, and also reduces nitrite to nitrogen.

$$HSO_3NH_2 + H_2O = NH_4HSO_4 \qquad \text{(II-25)}$$

$$HSO_3NH_2 + NO_2^- = N_2 + H_2O + SO_4^{--} + H^+ \qquad \text{(II-26)}$$

The latter reaction should be taken into account when passivating with sodium nitrite following cleaning with sulfamic acid; if too much acid is left in a unit, passivation will be incomplete. The subject of passivation is discussed in Chapter V.

HYDROFLUORIC ACID

Because of its extreme toxicity and the difficulty of safely handling its solutions, hydrofluoric acid is seldom used, as such. Instead, ammonium bifluoride, NH_4HF_2, is added to dilute hydrochloric acid to obtain the special effects associated with the chemistry of hydrofluoric acid. This subject is pursued further in Section II.5.

The stability of the hexafluoroferric ion, FeF_6^{-3}, can be exploited in eliminating ferric ion corrosion of steel. As FeF_6^{-3} does not oxidize iodide, the potential of the half-reaction

$$Fe^{++} + 6F^- = FeF_6^{-3} + e^- \qquad \text{(II-27)}$$

must be less negative than -0.536 V, the oxidation potential of the iodine-iodide half-reaction

$$3I^- = I_3^- + 2e^- \qquad \text{(II-28)}$$

Yorkgitis(13) has reported good results in cleaning predominantly iron oxide deposits from the hearth staves of blast furnaces and other cooling water systems in steel mills. Here too, the stability of FeF_6^{-3} is a significant factor in bringing oxides of iron into solution. The dissociation constant of the reaction

$$FeF_6^{-3} = Fe^{+3} + 6F^- \qquad \text{(II-29)}$$

is $K \cong 5 \times 10^{-16}$. As pointed out earlier, hydrofluoric acid is a weak acid ($K_a = 6.6 \times 10^{-4}$), so the effect of hydrogen ion itself is negligible in solubilizing oxides of iron. Solutions containing 5 percent of ammonium bifluoride and 1.0–1.5 percent of hydrofluoric acid are suggested by Yorkgitis(13) to solubilize deposits containing 60–90 percent iron oxides; if a significant concentration of calcium carbonate is present (8–10 percent), the amount of HF is increased. The inhibitor used is the hydrochloride of a rosin-amine ketone. Note that, in common with other halogen acids, HF is not recommended for use in stainless steel equipment.

II.2.b Organic Acids

In certain applications it is advisable to use solutions of acids in which the hydrogen ion activity is less than that of the mineral acids. These are moderately weak organic acids that, with the exception of formic acid, are less hazardous to handle and more easily disposed of than the strong mineral acids described in the previous section. The most commonly used organic acids are hydroxyacetic (glycolic), formic, citric, and less frequently, oxalic acids. Some of these owe much of whatever solvency they have to their ability to form complexes with cations, rather than to the activity of their hydrogen ions. Nevertheless, all of them must be inhibited before they are applied to industrial equipment. In general, these acids have much less capacity for dissolving oxides of iron than does hydrochloric acid, and they are incapable of dissolving deposits within pits. For these reasons, they are used only when hydrochloric acid is unsuitable, as for instance, in once-through utility boilers, in which pendant superheaters, or reheaters, are often fabricated of stainless steels, and also may be undrainable.

As these acids are expensive, they are applied at the lowest effective concentration and at the highest practicable temperature to hasten rates of solution. Although positive circulation at a minimum rate of 1 fps is essential, as with inorganic acids, excessively high velocity must be avoided, for this retards the formation of inhibiting films and rapidly erodes base metal. Cleaning with these agents is invariably unsatisfactory if minimum velocity and temperature cannot be maintained. This is a particularly important consideration in the case of steam generators, in which it is generally agreed that an incomplete chemical cleaning is much worse than no cleaning at all.

HYDROXYACETIC (GLYCOLIC) ACID

This organic acid is available commercially as a 70 percent aqueous solution. It is often used to clean stainless steels, especially in preboiler systems and once-through steam generators, and to remove water-formed scales from food processing and other industrial equipment; it functions by forming weak complex metal hydroxyacetates. For cleaning boilers hydroxyacetic acid at 2 percent is combined with formic acid at 1 percent, heated to 200-220°F, and circulated at a minimum rate of 1 fps for 4-6 h.

FORMIC ACID

Commercial aqueous solutions contain 88-90 percent of formic acid, which must be cautiously handled, for it produces dangerous skin burns and is percutaneously introduced into the human body. The reason for combining hydroxyacetic and formic acids is not apparent, as both form ferric comples of about the same stability, and they are of comparable strengths as acids.

$$Fe^{+3} + HOCH_2COOH = Fe(HOCH_2COO)^{++} + H^+ \quad K_1 = 6.3 \times 10^2 \quad \text{(II-30)}$$

$$Fe^{+3} + HCOOH = Fe(HCOO)^{++} + H^+ \quad K_1 = 7.9 \times 10^1 \quad \text{(II-31)}$$

$$HOCH_2COOH = HOCH_2COO^- + H^+ \quad K_a = 1.5 \times 10^{-4} \quad \text{(II-32)}$$

$$HCOOH = HCOO^- + H^+ \quad K_a = 1.8 \times 10^{-4} \quad \text{(II-33)}$$

This may be an instance of the frequently applied and strongly held principle that if one thing works well, several must work better. At any rate, if there is any advantage to combining these acids, it is not obvious from their chemical properties. Either acid, or combinations of the two, can be disposed of by incineration. The pH of a 2 percent solution of hydroxyacetic acid is 2.16; that of a 1 percent solution of formic acid is 2.51.

OXALIC ACID

This is the strongest of the organic acids considered here, although its solubility in water is limited to 8-9 percent. Solutions of the acid attack titanium, but are suitable for stainless steels, and they can be conveniently disposed of by incineration. Having no particular advantage over the other organic acids, oxalic acid is used only occasionally for chemical cleaning.

CITRIC ACID

This acid has been suggested for the preoperational cleaning of preboiler systems;[(4)] it is also useful for certain aluminum equipment, and is the chemical of choice for titanium heat exchangers. For the latter two applications, the usual concentration is 3 percent circulated at 150°F. For cleaning boilers and associated equipment, 3-5 percent citric acid is customarily neutralized with ammonia ("ammoniated") to a pH in the range of 3.5–4.0, then circulated at 2 fps for 6–8 h. The dissolution of iron oxides is slow, but once dissolved ferric and ferrous ions are sequestered by citrate in both acidic and alkaline solutions. Solutions of citric acid are essentially nontoxic, biodegradable (indeed, they ferment spontaneously if allowed to stand for any length of time), and are chloride-free, making them useful for application to austenitic stainless steels.

The remarkable stability of the citratoferrate (III) ion arises because three electrostatic bonds are supplemented in the complex by three dative covalent bonds. That is, the free tridentate ligand, citrate ion, in addition to three negative charges, contains three unshared electron pairs that are capable of functioning as Lewis bases. With transition metals, such as iron, these unshared electron pairs enter the empty *d* orbitals in the central metal ion, thus greatly increasing the stability of the complex, which in this instance might just as well be considered a weak salt.

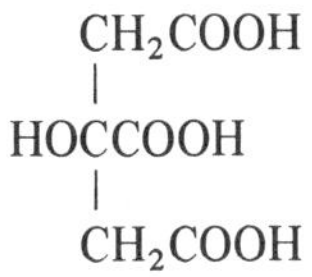

Citric Acid

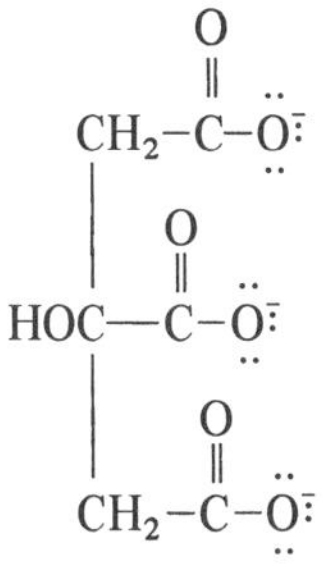

Citrate Ion

$$Fe^{+++}(C_6H_5O_6{}^{---}) = Fe^{+++} + C_6H_5O_6{}^{---} \qquad K = 1 \times 10^{-25} \qquad \text{(II-34)}$$

The significance of this small value of the instability constant, K, is made evident by a consideration of the solubility product of ferric hydroxide, from which the concentration of ferric ion in a solution at pH 10 can be calculated, for example.

$$(Fe^{+3})(OH^-)^3 = 6 \times 10^{-38}$$

At pH 10,

$$(OH^-) = 10^{-4} \text{ mol/liter}$$

$$(Fe^{+3}) = 6 \times 10^{-38}/(10^{-4})^3$$

$$= 6 \times 10^{-26} \text{ mol/liter}$$

The instability constant of the citratoferrate (III) complex, FeC, is

$$\frac{(Fe^{+3})(C^{-3})}{(FeC)} = 1 \times 10^{-25}$$

Suppose that an excess of citrate is present such that,

$$(C^{-3}) = 3(FeC)$$

Then,

$$\frac{(Fe^{+3})3(FeC)}{(FeC)} = 1 \times 10^{-25}$$

or,

$$(Fe^{+3}) = 1 \times 10^{-25}/3$$
$$= 3.3 \times 10^{-26} \text{ mol/liter}$$

Thus, as the concentration of ferric ion contributed by the citratoferrate complex is about one-half that contributed by ferric hydroxide at pH 10, the latter does not precipitate in the presence of a three-fold excess of citrate. This shows the importance of not allowing the concentration of citric acid to dwindle during the course of a cleaning operation for the dissolution of iron oxides, a point mentioned by Duncan.[(4)] Another procedure using alkaline citrate is discussed in Section II.6.

II.2.c Acidic Emulsions

As mentioned before, no acid can dissolve inorganic deposits that are coated with oil or other organic material, for which reason alkaline solvents were described in Section II.1 with which a preliminary deoiling can be accomplished. If only a light coating of oil is present, however, time can often be saved by introducing an acidic emulsion of 50–80 percent diesel fuel or aromatic hydrocarbon in one of the acidic solutions already discussed. For carbon steel, inhibited 5 percent hydrochloric acid is suggested; for stainless steels, 10 percent phosphoric acid might be used. These emulsions can be stabilized by adding 0.1 percent of an oil-soluble nonionic surfactant.

II.2.d Special Considerations in the Use of Acids

Before proceeding with a chemical cleaning it is essential to determine that the solvent chosen is compatible with the metallurgical materials of construction of the unit to be cleaned. There must be no guessing about this, as catastrophic damage can be done in a short time by an incompatible solvent; for this reason it is always wise to err on the side of conservatism in selecting acidic solutions, concentrations, and temperatures. Several of the more common metals and alloys used to fabricate industrial equipment are now considered.

STEELS

Carbon steel is by far the most commonly used metal of construction in the manufacturing industries. It is used in boilers, heat exchangers, waste heat steam generators, reactors, stripping columns, piping, and tankage, primarily because of its relatively low cost. It is highly vulnerable to corrosive attack, however, especially by hydrogen ion, hydrogen sulfide, and oxygen. Despite this, carbon steel is widely used in aqueous recirculating cooling systems, which when properly treated,[(14)] produce very low rates of corrosion. Alloying elements are added to carbon steel to improve particular properties. A few percent of chromium, for instance, increases corrosion resistance. Molybdenum, added as a toughener to low-carbon steel at a concentration of about 1 percent, doubles the tensile strength of the steel at elevated temperatures, making it more suitable for steam service.

CAST IRON

This material, which is more readily corroded by strong acids than is carbon steel, is encountered in pump casings, water jackets, and in bubble caps in tray columns. It can be safely cleaned with inhibited hydrochloric acid at 5-10 percent at not more than 140°F. As most cast iron objects are massive, the increased corrosion rate as compared with mild steel is not a major concern. Because cast iron is to some extent porous, it is advisable to extend the usual neutralization treatment by an hour or so. Cast iron water jackets and columns fitted with cast iron bubble caps have been routiely cleaned with 10 percent hydrochloric acid with no evidence of deleterious effects. In all cases, however, the temperature is kept below 140°F, and inhibitor is added at twice the manufacturer's recommended dosage.

COPPER ALLOYS

Alloys are solid solutions of two or more chemical elements formulated for special purposes. Those of copper are used for heat exchanger tubing, surface condensers, stage heaters, and evaporators because of their high resistance to corrosion in aqueous environments. Copper itself is quite resistant to corrosion and has high thermal conductivity, but as it is not very strong, it is usually alloyed with other elements, notably zinc, tin, and aluminum. High velocity or turbulent flow of water strips the normally protective film of cuprous oxide from the surface of copper, making it susceptible to corrosion by air, oxidizing agents, and aerated solutions, such as cooling water, containing ammonia or ammonium salts. When copper is combined with zinc it produces a series of alloys called brasses that are resistant to such impingement. Muntz metal is a brass that is useful in fresh-water service, whereas admiralty brass, which contains about 1 percent of tin as well as zinc, is resistant to sea water. Aluminum brass contains about 2 percent of aluminum and something less than 0.1 percent of arsenic; it is frequently used to fabricate condenser tubes for service in brackish or polluted waters. When tin is substituted for zinc the resulting alloys are called bronzes, which are characterized by high tensile strength. Aluminum bronze contains 90 percent of copper and 10 percent of aluminum. Other common copper alloys are cupro-nickel, which is used in feed water preheaters, and monel, which being highly resistant to hydrofluoric acid, is often used in HF-alkylation plants.

ALUMINUM AND ZINC

Aluminum is used for certain specialized tankage, but is seldom encountered in manufacturing facilities. It is attacked by most alkalies and acids, with the exception of citric acid, which at 3 percent and 150°F is an acceptable cleaning solution for aluminum. If properly inhibited, galvanized iron and other zinc surfaces can be successfully cleaned with 5 percent hydrochloric acid.

STAINLESS STEELS

When iron is combined with chromium and nickel a series of alloys is obtained that shows high resistance to corrosion by acids. The most widely used stainless steel contains 18 percent chromium and 8 percent nickel. As these alloys are subject to stress corrosion cracking in the presence of chloride ion, they are unsuitable for cooling water service, although for other reasons they are frequently used in this application. Stainless steels are attacked by formic, oxalic, hydrochloric, and hydrofluoric acids, and also by ferric and cupric ions, as well as sea water.

TITANIUM

Titanium is lighter and stronger than steel and is resistant to corrosion by ammonia, hydrogen sulfide, and to impingement attack by salt water. Because of the possibility of hydrogen embrittlement, it is advisable to be cautious in using mineral acids to clean titanium; citric acid appears to be the best choice, at present.

Table II-8 is a succinct summary of the many pieces of information included in this section on the application of acids for chemical cleaning. The following list of metals is referred to by numbers under the headings *Compatible* and *Incompatible* in Table II-8. The column headed pK_1 is a list of values of the first (or only) dissociation constant of each acid and is a measure of the strength of the acid: the larger the number, the weaker the acid.

1. Carbon steel
2. Carbon steel + 1% molybdenum
3. Carbon steel $<$ 5% chromium
4. Cast iron
5. Admiralty brass
6. Aluminum brass
7. Muntz metal
8. Aluminum bronze
9. Cupro-nickel
10. Monel
11. Copper
12. Aluminum
13. Zinc (or galvanizing)
14. Stainless steels
15. Titanium

A special proprietary inhibitor is essential for successfully cleaning zinc with acid; this is discussed in Section II.3, below.

In concluding this preliminary discussion on acid cleaning, it is worth pointing out that acid solutions should not be introduced into systems in which steel is connected to copper alloys. In such a coupled system the steel, in effect, becomes a sacrificial anode, and dissolves rapidly in aerated solutions. Furthermore, any copper brought into solution electrolytically plates on steel surfaces, intensifying corrosion of the latter.

II.3 CORROSION INHIBITORS

The corrosion rate of carbon steel in 10 percent hydrochloric acid at 160°F is approximately 0.6 mils/h. Therefore, before applying acidic cleaning solvents to industrial equipment, corrosion inhibitors must be added to the acid solution that can prevent metal from being attacked without interfering with the dissolution of fouling material. With the

TABLE II-8

Properties of Acids Used for Chemical Cleaning

Acid						Metals and alloys	
Name	Formula	%	lb/gal	pK_1*	Normality	Compatible	Incompatible
Hydrochloric	HCl	5	0.427	−6.1	1.40	1–11	12–15
Hydrochloric	HCl	7.5	0.651	−6.1	2.13	1–11	12–15
Hydrochloric	HCl	10	0.874	−6.1	2.87	1–11	12–15
Sulfuric	H_2SO_4	5	0.430	−3.0	1.05	1–11, 14	12, 13, 15
Sulfuric	H_2SO_4	10	0.890	−3.0	2.17	1–11, 14	12, 13, 15
Sulfuric	H_2SO_4	15	1.379	−3.0	3.37	1–11, 14	12, 13, 15
Phosphoric	H_3PO_4	3	0.254	2.21	0.045	1–11, 14, 15	12, 13
Phosphoric	H_3PO_4	5	0.428	2.21	0.060	1–11, 14, 15	12, 13
Phosphoric	H_3PO_4	10	0.0879	2.21	0.096	1–11, 14, 15	12, 13
Sulfamic	HSO_3NH_2	10	0.879	0.99	1.03	1–3, 5–11, 14	4, 12, 13, 15
Hydrofluoric	HF	1	0.084	3.18	0.018	1–3, 10	12–15
Hydroxyacetic	$HOCH_2COOH$	2	0.163	3.82	0.007	1–11, 14	12, 13, 15
Formic	$HCOOH$	1	0.084	3.75	0.003	1–11, 14	12, 13, 15
Citric	$C_6H_8O_7$	3	0.250	3.13	0.010	1–12, 14, 15	13
Citric + NH_3	$C_6H_8O_7$	3	0.250	3.13	0.0003	1–4, 14, 15	5–13
Oxalic	$C_2H_4O_4$	3	0.253	1.27	0.110	1–12, 14	13
Chromic	CrO_3	10	0.898	−0.98	—	1–4, 14, 15	5–13

* $pK_1 = -\log_{10} K_1$

exception of the now infrequent use of arsenite, antimony salts, stannous salts, and halides, all inhibitors used in chemical cleaning processes are organic compounds or mixtures thereof. The most effective of these contain nitrogen, sulfur, or oxygen, and function by adsorption, chemisorption, or complex formation on the metallic surface to be protected. The thickness of the protective film thus formed ranges from monomolecular to several hundred angstroms. The tenacity of the surface layer varies depending upon the metal, the inhibitor, concentration and the velocity of the cleaning solution, and its temperature. Some particulars relating to corrosion inhibitors are examined in the following subsections.

II.3.a Mechanism of Corrosion Inhibition

The most effective corrosion inhibitors are capable of forming chemisorption bonds with metal. This capacity arises from the presence of unshared electron pairs that enter incomplete *d* orbitals in transition metals, e.g., iron, nickel, chromium. Zinc and aluminum, neither of which is a transition metal, are difficult to inhibit because chemisorption bonds cannot form on the surfaces of these metals. Thiourea is an example of a corrosion inhibitor containing several unshared electron pairs. Nitrogen, being in Group V of the periodic system, contains five electrons in its outer shell, three of which are engaged in covalent bonding with hydrogen and carbon. Sulfur, in Group VI, contains six valence electrons, of which only two are used for covalent bonding to carbon in thiourea.

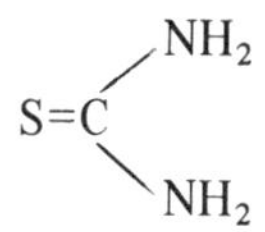

Thus, thiourea is well equipped to form chemisorption bonds with metals containing incomplete *d* orbitals, and would be expected to be an efficient corrosion inhibitor for transition elements, which it is.

Because of the presence of unshared electron pairs, thiourea functions as a Lewis base by accepting hydrogen ions.

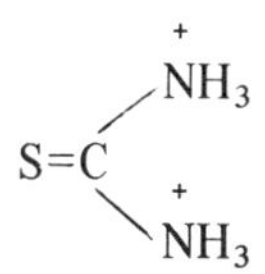

On account of the acquired positive charge, the molecule tends to migrate to cathodic areas, and is thus sometimes classified as a cathodic inhibitor. Hoar and Holliday[15] have shown, however, that thiourea and its derivatives inhibit both cathodic and anodic reactions, being more cathodic at low concentrations and more anodic at higher concentrations.

The surfaces of most metals tend to carry a slight positive charge, and therefore, attract molecules carrying negative charges, as well as polar organic compounds. Other molecules become attached to metallic surfaces through the agency of weak Van der Waals forces. Such attachments are characterized by low heats of adsorption, whereas chemisorption bonds show much higher values, in the range, for example, of 3000–8500 calories/mole for substituted quinolines on mild steel in 5 percent sulfuric acid.[15] As a rule, the more effective organic inhibitors are capable of forming chemisorption bonds.

Trabanelli and Carassiti[16] have mentioned a third type of inhibition in which the inhibitor forms a stable complex on the metallic surface. Quinoline, for instance, forms a complex with ferrous and chloride ions in hydrochloric acid that is attached to steel surfaces as an insoluble film. Here too, the unshared pair of electrons on the heterocyclic nitrogen atom plays a role.

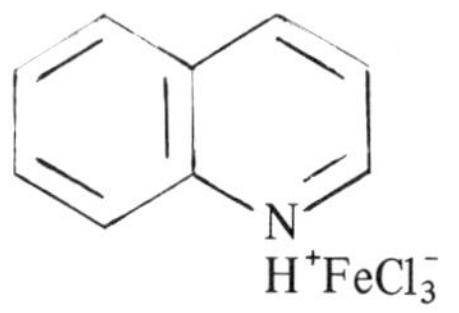

Unshared electrons can be made more accessible for forming chemisorption bond by taking advantage of the inductive effect[17] of alkyl groups introduced at appropriate positions in heterocyclic nitrogen compounds. Ayers and Hackerman,[18] in studying the inhibition of hydrochloric acid by methylpyridines, found that inhibition was greatest with 2- and 4-picoline, lower with 3-picoline, and least with pyridine, which correlates with the electron densities on the nitrogen atom in each compound. Thus, there is the possibility of tailoring a chemical compound to maximize its corrosion inhibiting capability.

II.3.b. Types of Inhibitors

Trabanelli and Carassiti[16] list over 150 structural formulas of organic compounds that have been proposed as corrosion inhibitors; a limited number of these warrant discussion here. For convenience, these are classified as cathodic (those that impede the reduction of hydrogen ion); anodic (those that limit the oxidation of iron); and general adsorption (those that form a physical barrier over the entire metallic surface). Cathodic inhibitors raise the hydrogen over-voltage, causing the corrosion potential of the protected metal to become more positive, for which reason they are sometimes called passivating inhibitors. They are suitable for inhibiting sulfuric and phosphoric acids, for example, but not hydrochloric or other halogen acids.(18) Anodic and general adsorption inhibitors are nonpassivating and, therefore, are appropriate for the halogen acids. Table II-9 contains listings of several useful corrosion inhibitors classified as just described.

TABLE II-9

Classification of Selected Corrosion Inhibitors

Cathodic	*Anodic*	*General adsorption*
Alkylamines	Benzotriazole (H)	Acetylenic alcohols
Benzotriazole (L)	Mannich bases	Benzotriazole
Diphenylamine	Pyridines	Dibenzylsulfoxide
Thioureas (L)	Quinolines	Diphenylamine
	Thiols	Furfuraldehyde
	Thioureas (H)	

L = at low concentration
H = at high concentration

As noted in the table, the mode of function of a couple of the inhibitors depends upon concentration. Benzotriazole, which is a specific corrosion inhibitor for copper and its alloys, has been studied by Mayanna and Setty,(19) who found that the dissolution rate of copper in N/10 sulfuric acid is a function of temperature, crystal structure, and the concentration of benzotriazole. The compound forms a tenacious, yellowish-white film on the surface of the copper; maximum inhibition was obtained at about 0.09 percent benzotriazole.

In general, inhibitors containing both nitrogen and sulfur in the same molecule show the highest efficiency as corrosion inhibitors. Thiourea and its derivatives are such compounds, as are thiomorpholine and phenothiazine, all of which are, or have been, protected by patents.(20,21,22)

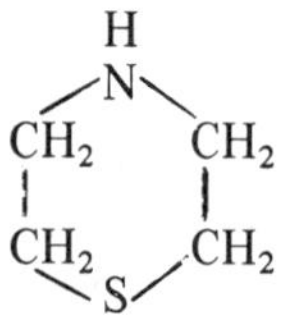

Thiomorpholine

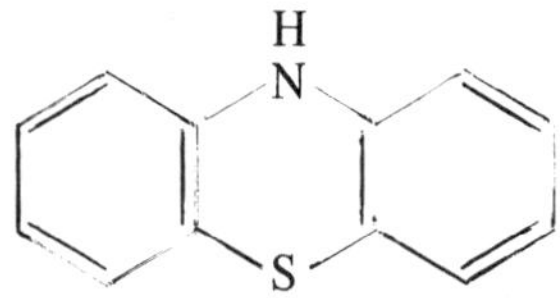

Phenothiazine

Substituted thioureas, although very efficient inhibitors of iron corrosion in acid solutions, facilitate the penetration of hydrogen into the metal, which can lead to embrittlement, particularly in steels of high tensile strength. The =C=S group is partially reduced at cathodic sites to H_2S, which also enhances hydrogen penetration, and in addition, attacks the iron because of the high oxidation potential of the half-reaction:

$$Fe + H_2S = FeS + H_2 \tag{II-35}$$

Thioureas tend to stimulate the dissolution of iron at low concentrations, so it is necessary to ensure that an adequate dosage is applied when using these compounds.

Alkylamines, which contain only one nitrogen atom, including nonylamine and decylamine, are not very efficient inhibitors for iron, nor are aniline or triphenylamine. Alkylamines have the further disadvantage of intensifying acidic attack on copper, brass, bronze, and zinc; copper and its alloys can be protected by aniline, pyridine, or cyclohexylamine. Oddly enough, cyclic polymethyleneimines,(23) which also contain a single nitrogen atom, are fairly efficient inhibitors, especially nonamethyleneimine and decamethyleneimine.

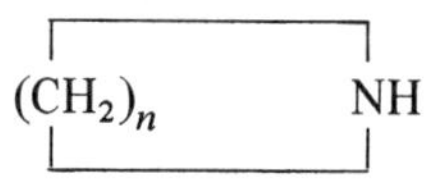

Polymethyleneimine

The efficiencies of compounds such as quinoline and its derivatives, that form complex protective layers composed of inhibitor, metal, and anion are significantly influenced by the anion in the acidic cleaning solution. For instance, Uhlig(8) quotes a corrosion rate of 260 mdd in 2*N* H_2SO_4, but only 48 mdd in 2*N* HCl, both inhibited with quinoline at 20°C. As the adsorption of cationic inhibitors on iron surfaces in sulfuric acid is weak, inhibition is poor in that acid. Other idiosyncracies can be attributed to the condition of the metallic surface being inhibited. Thus, Trabanelli and Carassiti[16] report that dibenzylsulfoxide is more firmly adsorbed on a cold-worked steel surface than on one that has been annealed. This inhibitor is most efficient in sulfuric acid; it is reduced in acidic solutions to the sulfide, which deposits as an insoluble protective film on metallic surfaces.

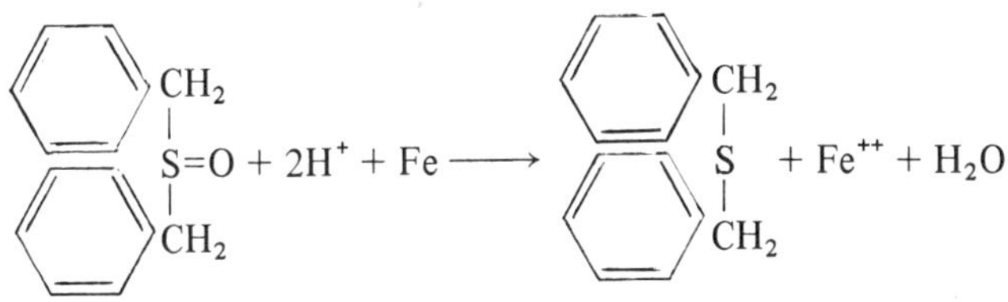

(II-36)

Benzylsulfoxide *Benzylsulfide*

Furfuraldehyde provides exceptionally good protection for copper surfaces in hydrochloric acid by polymerizing to a resinous film. Propargyl alcohol, $HOCH_2C{\equiv}CH$, is widely used to protect metal surfaces, which it does by polymerizing, in use, to an unsaturated ether that forms a physical barrier 20–200 Å thick. Although, aside from skin irritation, there is seldom any special hazard in handling acetylenic inhibitors, Crouse[24] has reported an instance of the detonation of a 30 percent aqueous solution of propargyl alcohol in aqueous sulfuric acid, upon being heated to 70°C. Chances are remote that such an event would occur with the dilute solutions handled in chemical cleaning, but it is well to be aware of the possibility, as inhibitors are often added to relatively concentrated acids.

Other acetylenic alcohols and derivatives have also been proposed,[25] usually applied in combination with mixtures of polar nitrogen compounds. Following are structural formulas of two of many acetylenic derivatives suggested as general adsorption corrosion inhibitors.

$$(C_2H_5)_2N{-}CH_2{-}C{\equiv}CH$$

3-diethylamino-1-propyne

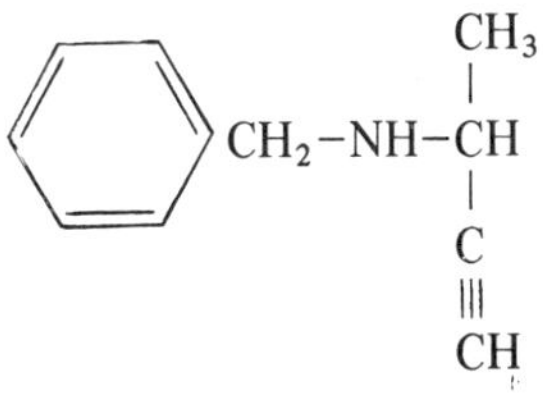

3-benzylamino-1-butyne

Mannich bases[26] are condensation products of the methylamination reaction between a primary or secondary amine, an α-ketone, and formaldehyde. Certain of these bases, in combination with thiourea, have been proposed as corrosion inhibitors for steel in hydrofluoric acid.[22] As an example of the Mannich reaction, the condensation of acetone, formaldehyde, and abietylamine is considered. When the resinous exudate of pine trees is distilled, the distillate is turpentine and the residue is rosin, which contains 80–90 percent of abietic acid and its anhydride.

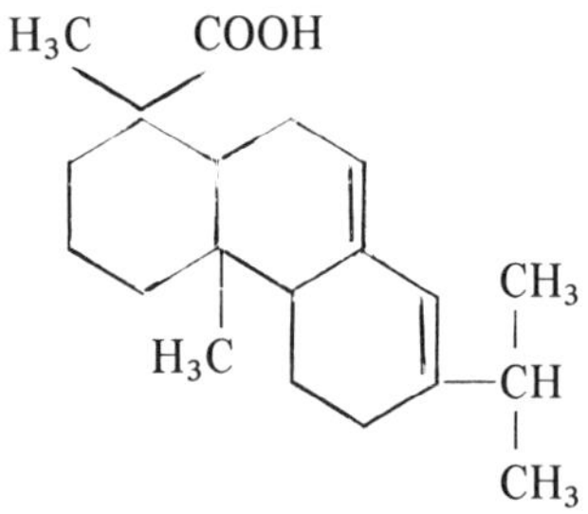

Abietic acid

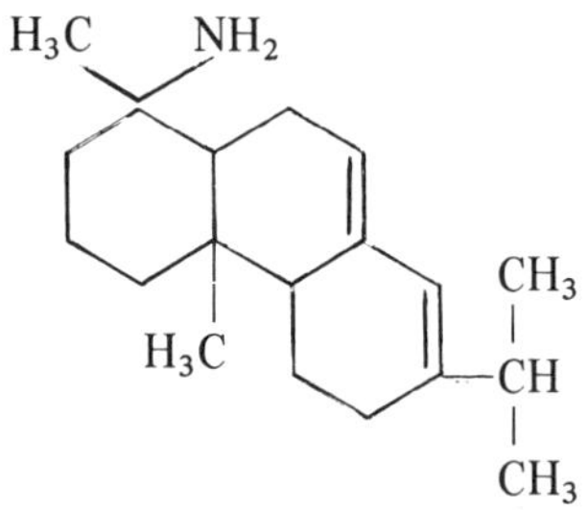

Abietylamine

Abietic acid can be converted to abietylamine by pyrolyzing its ammonium salt, then applying the Hofmann reaction to the resulting amide. If the abietyl radical is represented by R, the reactions are as follows:

$$RCOONH_4 \rightarrow RCONH_2 + H_2O \qquad \text{(II-37)}$$

$$RCONH_2 + NaOBr \rightarrow RNH_2 + NaBr + CO_2 \qquad \text{(II-38)}$$

Having obtained abietylamine, the Mannich reaction is carried out; using acetone, formaldehyde, and abietylamine as an example, the product is abietylaminodibutanone, a Mannich base.

$$2CH_3COCH_3 + 2HCHO + RNH_2 \rightarrow RN(CH_3COCH_2CH_2)_2 + 2H_2O \qquad \text{(II-39)}$$

Other inhibitors proposed for use with hydrofluoric acid are dibenzylsulfoxide or pyridiniumxanthogenate, with thiourea added to restrain pitting. Vosta, et al.[27] claim that these combinations reduced the corrosion rate of carbon steel from 2500 mpy to 30 mpy in 1 percent HF at 140°F.

Mansfield and Kenkel[28] have found that organic compounds containing nitro groups are effective for inhibiting the corrosion of titanium alloys in hydrochloric acid. Nitro groups in picric acid, p-nitroaniline, p-nitrophenol, nitrobenzene, and 2,4-dinitrophenol are apparently reduced on the titanium surface causing polarization to a more positive potential. The information contained in this subsection is summarized in Table II-10.

TABLE II-10

Appropriate Inhibitors for Chemical Cleaning

Metal	*Acid*	*Inhibitor*
Carbon steels	HCl	Thioureas, dibenzylsulfoxide, quinolines, propargyl alcohol.
	HF	Mannich bases, thioureas, dibenzylsulfoxide.
	H_2SO_4, H_3PO_4	Thioureas, dibenzylsulfoxide.
Stainless steels	H_2SO_4, H_3PO_4, HSO_3NH_2, Organic acids*	Trialkylamines, diphenylamines, thioureas, dibenzylsulfoxide, quinolines.
Copper alloys	H_2SO_4, H_3PO_4, HCl	Cyclohexylamines, aniline, pyridine, furfuraldehyde.
	Organic acids*	Thioureas, dibenzylsulfoxide.
Titanium	HCl	Picric acid, p-nitroaniline, p-nitrophenol.
	Citric acid	(Neutralized to pH 4 with aqueous ammonia.)

* Oxalic, citric, hydroxyacetic, and formic acids.

Note that Table II-10 has nothing to say about the important factors of temperatures, concentrations, and duration of treatments, these are taken up next.

II.3.c Commercial Formulations and Applications

As in the marketing of water-treating formulations, a few commonly used corrosion inhibitors are cloaked under a spate of tradenames and numbers. Nevertheless, it is most unlikely that any chemical cleaning contractor would find it worth his while to attempt to formulate the chemicals discussed above for industrial use. Rather, it is customary to purchase commercial formulations and use these as recommended by their manufacturers, paying scrupulous attention to limitations on temperatures and acid and inhibitor concentrations. The material in the preceding subsection is provided for the information of clients; it has little practical use in contracting or carrying out an actual chemical cleaning.

Several chemical companies have developed safe, efficient formulations for use with specific acids and metals, a number of which were evaluated some years ago by Technical Group T-8 of the National Association of Corrosion Engineers.[29] The following tradenames, most of them encompassing several different formulations designated by identifying numbers, were tested: O'B-Hibit,* Armohib,** Dowel,† Cronox,†† and Rodine.††† Twelve of these inhibitors were examined to determine the effects on corrosion rates of the time of exposure, temperature, presence of sulfide and ferric ions, concentrations of acids and inhibitors, and galvanic coupling. In addition, an appendix is included containing the recommendations of the manufacturers for the use of these materials. This information is a great deal more valuable, incidentally, than the numerous papers reporting results with "Inhibitors A, B, and C," not otherwise identified by name, a form of advertising that for some reason continues to be published as if it were technical material.

Commercial inhibitors invariably are mixtures, some of them by-products of manufacturing processes containing unknown active ingredients; the application of such products is necessarily empirical. Other additives that are often included in commercial formulations are surfactants and vapor phase inhibitors. The former increase the wettability of metal surfaces by lowering interfacial tension, thereby facilitating the orientation of the general adsorption type of inhibitor. Vapor phase inhibitors, e.g., cyclohexylamine, which have high vapor pressures, volatilize from warm cleaning solutions, then condense on unprotected metal above the solution (in the upper part of a steam drum in a boiler, for instance), and prevent acid vapor from attacking the metal. This may seem an uncertain method, but inhibitors are available that can adequately control corrosion, even in

---

* O'Brien Industries, Inc., 95-T Dorsa Avenue, Livingston, N.J. 07039

** Armour and Company, Greyhound Tower, Phoenix, Arizona 85077

† Dowell Division, Dow Chemical Company, Houston, Texas 77027

†† Aquaness Chemical Co., 2434 Holmes Road, Houston, Texas 77051

††† Amchem Products, Inc., Brookside Ave., P.O. Box 33, Ambler, Pa. 19002

such drastic procedures as injecting hydrochloric acid into steam to clean pipelines.

At excessively high temperatures some inhibitors decompose, polymerize, or are otherwise altered and lose their effectiveness, while too high a recirculation rate (> 2 fps) of acidic cleaning solutions invariably stimulates corrosion. Then too, the corrosion rate of steel in mineral acids increases with carbon content; high carbon steel (C = 0.3% +) and cast iron are always difficult to inhibit. Galvanized iron and other zinc-containing material can be safely cleaned with 5 percent hydrochloric acid at 150°F if the acid is inhibited with 1.25 percent of Cronox 260, a film-forming inhibitor. In general, it is unwise to inhibit inventory acid, that is, concentrated acid kept in storage for later use, as some of the amine inhibitors polymerize and precipitate. Conservative cleaning contractors add inhibitors to acids at double the volume recommended by the manufacturers; for cleaning carbon steel or cast iron, hydrochloric acid is inhibited at four times the recommended volume. A few inhibitors are dyed to show their presence, but if there is the slightest doubt about the state of an acidic cleaning solution, the steel wool test described in Chapter XI should be made.

II.4 WETTING AGENTS

The results of acid cleaning are usually improved by including 0.1–0.2 percent of a surfactant in the solvent to serve as a scouring and emulsifying agent. It may well be that the inorganic portion of a deposit is covered or enclosed by oily material, grease, or the like, shielding it from the solubilizing action of the acid. As with the inhibitors, there are many surface active products to choose from, of which only three need be mentioned here. The most satisfactory wetting agents for chemical cleaning are the nonionic types exemplified by the series of alkylphenol–ethylene oxide condensation products marketed under the tradename Igepal.* These are available either as water-soluble or oil-soluble products, their solubility being determined by the number of ethoxy groups in the molecule; the larger the value of n in Eq. (II-40), the greater is the solubility in water. A typical reaction for preparing this type of compound is the condensation of nonylphenol with ethylene oxide to yield nonylphenoxypolyoxyethylene ethanol.

---

* Manufactured by GAF Corporation, 140 W. 51st Street, New York, N.Y. 10020

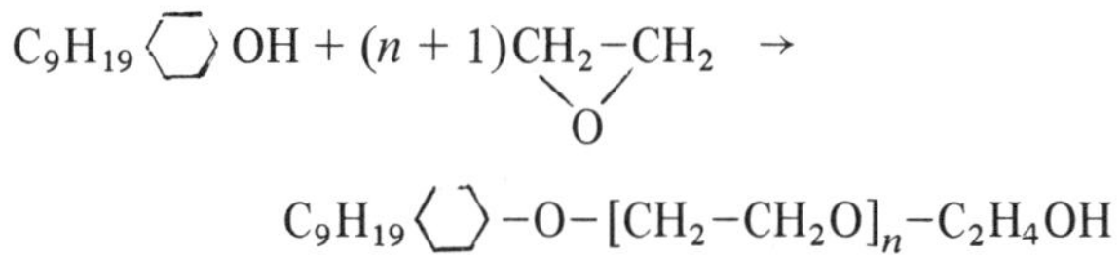

(II-40)

As these products are sensitive to heat, they must be limited to use at 150°F, or less. At temperatures up to 200°F anionic surfactants can be used although, in general, the nonionic products perform better. Sodium dioctylsulfosuccinate, marketed under the tradename Aerosol OT* has been used for many years as an emulsifying agent for oil, its only disadvantage being a propensity to foam.

$CH_2COOC_8H_{17}$
|
$CHCOOC_8H_{17}$
|
SO_3Na

Sodium dioctylsulfosuccinate

Steinhauer(30) has disclosed a method for preparing anionic surfactants by alkylating diphenylether with dodecylbromide, then sulfonating the alkylated ether with sulfur trioxide in methylene chloride. The final product is dodecylphenoxybenzenedisulfonic acid, which is marketed as the disodium salt under the tradename Dowfax.† Feighner(31) has very thoroughly described the chemistry of the alkylsulfonates, observing that at higher molecular weights, i.e., $C_{16}+$ in the alkyl group, the compounds tend to adsorb on metallic surfaces and prevent wetting; this implies a certain degree of inhibiting effect with these chemicals. Compounds with short alkyl chains are the best wetting agents; those with linear alkyl groups are much more easily biodegraded than those with branched chains. Alkaline deoiling, as described in Section II.1, can often be made more efficient by adding 0.1-0.2 percent of a wetting agent, although some of them hydrolyze readily in hot alkaline solutions.

* Manufactured by American Cyanamid Co., Process Chemicals Dept., Wayne, N.J. 07470

† Manufactured by Dow Chemical Co., 2020 Dow Center, Midland, Michigan 48640

II.5 SPECIAL AGENTS

There are a number of chemicals that are used under particular circumstances to enhance or supplement the effect of the cleaning solvents already discussed. Ammonium bifluoride, for instance, is added to hydrochloric acid and other acids to dissolve silicates in water-formed deposits. Stannous chloride is used with hydrochloric acid to reduce ferric ion brought into solution by the dissolution of rust and magnetite arising from the corrosion of steel. Gluconic acid is useful for bringing into solution certain resistant iron salts and oxides. Thiourea, substituted thiourea, sodium bromate, ammonium persulfate, and sodium nitrite are all used in solutions devised to dissolve and keep in solution metallic copper deposited on steel surfaces, most often in boilers. Each of these chemicals is discussed individually in the following pages.

II.5.a Ammonium Bifluoride

When ammonium bifluoride, NH_4HF_2, is dissolved in hydrochloric or other acid, hydrofluoric acid is formed because of its small acidic dissociation constant, $K_a = 6.6 \times 10^{-4}$. This acid specifically broaches water-formed silicate scales such as acmite, $Na_2O \cdot Fe_2O_3 \cdot 4SiO_2$, analcite, $Na_2O \cdot Al_2O_3 \cdot 4SiO_2$, and hydrated magnesium orthodisilicate, $Mg_3Si_2O_7 \cdot 2H_2O$, with the formation of fluosilicic acid and hexafluoroferric or hexafluoroaluminum ion.

$$Na_2O \cdot Al_2O_3 \cdot 4SiO_2 + 36HF = 2AlF_6^{-3} + 4H_2SiF_6 + 2Na^+ + 4H^+ + 12H_2O \qquad \text{(II-41)}$$

The addition of NH_4HF_2 to citric and other acids increases the rate of solution of magnetite and ferric oxide through the formation of the hexafluoroferric ion, which has a very small dissociation constant.

$$FeF_6^{-3} = Fe^{+3} + 6F^- \qquad K \cong 5 \times 10^{-16} \qquad \text{(II-42)}$$

Also, the sequestration of ferric ions by fluoride lowers the oxidation potential of the following reaction, minimizing additional corrosion by the solvent.

$$2Fe^{+3} + Fe = 3Fe^{++} \qquad \text{(II-43)}$$

In measuring the corrosivity of NH_4HF_2 solutions, Poff[(32)] found that in phosphoric, sulfamic, and citric acids, all at 5 percent, ammonium bifluoride became less corrosive to mild steel as its concentration was increased from 0 to 5 percent. He attributed this to an inhibitive effect of the ammonium ion, but it appears more likely that it is caused by the decrease in hydrogen ion activity attributable to the association of hydrofluoric acid. Thus, the reaction of free hydrogen ions from whichever acid is used, with fluoride ions from NH_4HF_2 to form undissociated HF, reduces the effective concentration of hydrogen ion significantly. In any event, corrosion rates in all of the acids tested were negligible, ranging from 15 to 65 mpy. The order of increasing corrosivity for the four acids tested was citric $<$ sulfamic $<$ phosphoric $<$ hydrochloric; the inhibitors, unidentified in Poff's work, also affect the corrosion rates.

Hydrofluoric acid and its salts should not be used for cleaning stainless steels as there is a possibility of the fluoride ion initiating stress corrosion cracking in these alloys. Furthermore, acidic solutions of fluoride salts are hazardous to handle for they are intensely corrosive to the skin and likely to be fatal, if ingested. The devastating effects of hydrogen fluoride on the human body has been graphically described by Mayer and Guelich[(33)] in a clinical study of several accident victims. In the interest of safety, the concentration of NH_4HF_2 is usually limited to 1.5 percent in cleaning solutions.

II.5.b Stannous Chloride

Although the extent of corrosion by hydrogen ion is negligible in a properly inhibited acid used at the specified temperature, ferric ion, released by acids from rust or magnetite, can cause extensive corrosion of steel.

$$2Fe^{+3} + Fe = 3Fe^{++} \qquad E^\circ = 1.21\text{ V} \qquad \text{(II-44)}$$

In an effort to eliminate the corrosive effect of ferric ion reagents have been proposed that reduce its activity. Fluoride, mentioned in Section II.5.a, is one of these; an alternative is to add a reducing agent. If fouling is present in which both ferrous sulfide and ferric oxide are found, hydrogen sulfide released by acid rapidly reduces ferric ion also released by acid. Note, however, that the reaction between acids and ferrous sulfide is normally very rapid, while that between ferric oxide and acids may be much slower. Thus, a considerable volume of hydrogen sulfide may be released that must be scrubbed or flared, regardless of how much Fe_2O_3 may be present. The net reaction of a combination of FeS, Fe_2O_3, and H^+ is represented by Eq. (II-45), but it must be borne in mind that some H_2S is always released from this sort of deposit.

$$FeS + Fe_2O_3 + 6H^+ = 3Fe^{++} + S + 3H_2O \qquad \text{(II-45)}$$

Although stannous chloride is often added to hydrochloric acid to reduce ferric ion, there are some deficiencies in the procedure. In cleaning a boiler, or other large vessel, the usual practice is to blend fill with dilute acid, pumping at such a rate that the boiler is filled within an hour or two. Then, as the acid circulates, stannous chloride (equivalent to 0.75–1.0 percent) is added to the supply tank on the chemical cleaning truck. By the time the stannous chloride enters the boiler a substantial amount of ferric ion corrosion will already have occurred, for which reason this expensive procedure is, to a great extent ineffectual. In addition, the amount of stannous chloride used is almost always insufficient for the stoichiometric requirement of Eq. (II-46).

$$Sn^{++} + 2Fe^{+3} = Sn^{+4} + 2Fe^{++} \qquad \text{(II-46)}$$

The commercial form of stannous chloride is $SnCl_2 \cdot 2H_2O$. A consideration of molecular weights and Eq. (II-46) shows that 2 lb of $SnCl_2 \cdot 2H_2O$ reduces 1 lb of ferric ion.

It is sometimes suggested that a cleaning solution should be drained and discarded when the concentration of ferric ion reaches some particular value, say, 0.5 percent. As steel tubes, shells, pumps, piping, and sample lines all react rapidly with Fe^{+3}, it is most unlikely that a significant amount of this ion will be found in a sample of cleaning solution, unless a heavy coating of rust or magnetite is initially present on the surfaces being cleaned. The attack on these surfaces by inhibited acids is uniform unless the solution has been overheated or circulated too rapidly, whereas attack by ferric ions usually takes the form of pitting in stressed areas, or the deepening of existing pits. Because ferric ion also oxidizes copper, nickel, zinc, and chromium, it is in most instances inadvisable to reuse a partially spent acidic solution for a second cleaning operation, although this is sometimes the basis for a low bid.

II.5.c Sodium Gluconate

The sodium salt of gluconic acid, $HOCH_2-(CHOH)_4-COOH$, produced by the fermentation of dextrose, is a powerful sequestrant of ferric and cupric ions in alkaline solutions. This suggests the possibility of simultaneously deoiling and derusting, which, indeed, can be done successfully in some instances. A hot solution (160-175°F) containing 5 percent of sodium hydroxide, 2 percent of sodium gluconate, and 0.2 percent of an anionic surfactant often proves to be effective in removing oily iron sulfide scales without the evolution of hydrogen sulfide. A pretreatment with alkaline gluconate may also open up certain fouling materials so that a subsequent treatment with an acidic cleaner is substantially more satisfactory than it would have been alone. An example of this is the exceptionally insoluble ferric ferrocyanide, $Fe_4[Fe(CN)_6]_3$, which is virtually unaffected by any acid unless first treated with alkaline gluconate. It is always worth trying this approach in solubility tests of refractory scales and deposits that resist more commonplace procedures. Further advantages of sodium gluconate are its nontoxicity (it is an approved food chemical), neutrality (pH 6.5-8.0 in concentrated solutions), and biodegradability. Gluconic acid, if used as such, must be handled in stainless steel or rubber-lined steel tanks; this is seldom done as the acid is quite weak, $K_a = 1.38 \times 10^{-4}$

At this point it is appropriate to present Table II-11, which contains a tabulation of logarithms of formation constants of complexes formed between some of the organic acids described in Section II.2.b, as well as those of ethylenediaminetetraacetic acid (EDTA) and the more common ionic components of scales and fouling deposits.

TABLE II-11

pK Values of Formation Constants of Organic Acids

Ion	*EDTA*	*Citric acid*	*Gluconic acid*	*Glycolic acid*	*Formic acid*
Ferrous	14.4	15.5	1.0	1.2	–
Ferric	25.1	25.0	37.2	2.8	1.9
Calcium	10.7	4.7	1.2	–	1.4
Magnesium	8.7	3.3	0.7	–	1.4
Cupric	18.8	14.2	19.6*	2.4	2.8
Zinc	16.6	11.4	1.7	3.3	2.0
Aluminum	16.1	20.0	–	1.4	1.8

* $[2\,Cu^{++} + 4OH^- + L^- = Cu_2(OH)_4L^-]$, where $L^- = HOCH_2\text{–}(CHOH)_4\text{-}COO^-$.

Note that the most stable complex in the table is that formed between ferric and gluconate ions. It is for this reason that sodium gluconate is often added to neutralizing and passivating solutions for boilers, where it prevents the precipitation of ferric hydroxide when acidic solutions are diluted and also removes any flash rusting that may develop because of delay in passivating after acid cleaning. It is obvious, too, from the values in Table II-11, that glycolic and formic acids could not be expected, as the result of complex formation, to be particularly effective for dissolving salts and oxides of the ions shown.

II.5.d Thiourea

Mention has already been made of thiourea and its derivatives as corrosion inhibitors; in addition, they are useful for removing metallic copper and its oxides from boilers. When copper is present in a boiler, galvanic corrosion in the form of severe pitting beneath a plate of metallic copper is almost always seen. Martin and Abel(34) have disclosed a method of using thiourea, or some of its derivatives, to sequester ionic copper so that it is not reduced by iron; their method has also been alluded to by Engle and Fitzgerald.(35)

When hydrochloric acid is added to cupric oxide it dissolves, then cupric ion is immediately reduced to a loose, easily displaced layer of copper on steel surfaces, at the same time solubilizing an equivalent amount of iron.

$$CuO + 2H^{+} = Cu^{++} + H_2O \qquad \text{(II-47)}$$

$$Cu^{++} + Fe = Cu + Fe^{++} \qquad \text{(II-48)}$$

The reaction in Eq. (II-48) can be blocked by adding thiourea to the hydrochloric acid solution to reduce cupric to cuprous ion, which then enters into a series of coordination complexes with excess thiourea. The latter, because of the presence of two amine groups, is more basic than an ordinary thioamide and is, thus, capable of forming mono salts with certain metallic cations. When using thiourea to complex cuprous ion, it is essential that an excess be provided to prevent the precipitation of the insoluble chloride mono salt. The sequence of possible reactions with copper ions is as follows:

Reduction of cupric ion by thiourea:

$$2Cu^{++} + CS(NH_2)_2 + H_2O = 2Cu^{+} + CO(NH_2)_2 + S + 2H^{+} \qquad \text{(II-49)}$$

Precipitation of insoluble thiourea mono salt (white):

$$Cu^+ + Cl^- + CS(NH_2)_2 = CuCS(NH_2)_2Cl \qquad \text{(II-50)}$$

Formation of soluble complex monomer:

$$Cu^+ + 3CS(NH_2)_2 = Cu[CS(NH_2)_2]_3{}^+ \qquad \text{(II-51)}$$

Dimerization of the monomer:

$$2Cu[CS(NH_2)_2]_3{}^+ = Cu_2[CS(NH_2)_2]_6{}^{++} \qquad \text{(II-52)}$$

Formation of a second monomer with a large excess of thiourea:

$$Cu^+ + 4CS(NH_2)_2 = Cu[CS(NH_2)_2]_4{}^+ \qquad \text{(II-53)}$$

Eq. (II-51) shows that a minimum of three moles of thiourea must be provided per mole of cuprous ion to keep the latter in solution, i.e., 3.6 lb of thiourea per pound of copper. Also, extra thiourea must be added to replace that oxidized to urea according to Eq. (II-49).

When magnetite in the presence of metallic copper is treated with a large excess of acid, copper is oxidized to cupric ion.

$$Fe_3O_4 + Cu + 8H^+ = 3Fe^{++} + Cu^{++} + 4H_2O \qquad \text{(II-54)}$$

If the acid solution contains sufficient thiourea, the otherwise unstable cuprous ion, formed as an intermediate product of the oxidation, is stabilized as the complex ions shown in the reactions of Eqs. (II-51), (II-52), and (II-53). In the absence of thiourea, however, any cuprous ion formed reverts to copper and cupric ion.

$$2Cu^+ = Cu^{++} + Cu \qquad \text{(II-55)}$$

In most instances of boiler fouling, the ratio of magnetite to copper is higher than the minimum demanded by Eq. (II-54).

A typical formulation for removing copper might contain 5.0 percent of hydrochloric acid, 2 percent of thiourea, and 0.4 percent of a proprietary corrosion inhibitor. Because of solubility considerations, the thiourea is dissolved separately in hot water, then added to the recirculating acid solution. After cleaning, a boiler must be carefully examined for loosened foils or tubes of metallic copper left after an inefficient cleaning, for copper rapidly deteriorates steel as the result of potential differences between the dissimilar metals. According to Atwood,[3] this single-phase treatment usually gives satisfactory results, especially if not more than 200 lb of copper is to be removed. Obviously, this method of chemical cleaning must not be used on copper alloys.

II.5.e Oxidizing Agents

Instead of maintaining copper in the cuprous state by adding a complexing agent, it is possible to stabilize the cupric state by adding an excess of some oxidizing agent. Nitrite, bromate, persulfate, and perborate have all been used. Cain and Schonhoff[36] have described a two-stage procedure for removing iron and copper from a boiler, in which 5 percent of inhibited hydrochloric acid is allowed to soak at 160–170°F for 6 h, then drained. The boiler is next rinsed with water at 140–150°F, after which ammonium bromate, NH_4BrO_3, is introduced at a concentration of about 0.5 percent. After 4 h the boiler is drained and rinsed. This sort of procedure is appropriate when the amount of copper to be handled is not known.

Wackenhuth and Richards[37] have investigated a three-phase treatment for removing copper, which they found to be so harsh that it produced deep pitting and gouging. In this method the boiler to be cleaned is first filled with a 5 percent solution of inhibited hydrochloric acid at 150°F, then allowed to soak for 6 h. The acid solution is drained under nitrogen, then the boiler is filled with an ammoniacal solution containing 1 lb of sodium bromate per pound of copper to be removed. An excess of the oxidizing agent must be present to keep copper in the cupric state so that the reaction represented by Eq. (II-48) cannot occur.

$$BrO_3^- + 3Cu + 12NH_3 + 3H_2O = Br^- + 3Cu(NH_3)_4^{++} + 6OH^- \quad \text{(II-56)}$$

The sodium bromate solution is heated to 150°F, allowed to soak for 6 h, then drained. Ammonium persulfate, $(NH_4)_2S_2O_8$, has been used as the oxidizing agent in this second step, but it too causes excessive pitting. Finally, a 5 percent solution of inhibited hydrochloric acid containing 0.75 percent of thiourea is heated to 150°F, and introduced into the boiler. After soaking for 6 h at 150°F, the solution is drained under nitrogen.

The use of sodium nitrite as an oxidizing agent in the removal of copper is discussed in the following section.

II.6 PATENTED CLEANING PROCESSES

The chemical cleaning methods described in this section are best left to experienced contractors who have proper equipment and who have made licensing arrangements for the patented procedures. As with any purchased service, however, it is to the purchaser's advantage to have a good understanding of the work to be accomplished and the methods to be used. Methods discussed here are ammoniated citrate and alkaline EDTA treatments for dissolving iron oxides and copper, the application of formaldehyde in the presence of sulfides to eliminate hydrogen sulfide, and a combination of ethylenediamine and hydroxylamine hydrosulfate that is useful for removing copper and its oxides from turbine blades.

II.6.a Ammonium Citrate

Alfano[38,39,40,41] has devised a method using ammoniated citric acid to dissolve iron oxides and copper that is especially useful for utility steam generators; the procedure is known commercially as the Citrosolv Process.* Because of the large formation constants of the citrate complexes of copper and iron (see Table II-11), oxides of the former are dissolved and held in solution at pH 4, while the latter, upon raising the pH to 10 and adding an oxidizing agent, is also brought into stable solution. In order to accomplish a chemical cleaning in a reasonable length of time, it is essential that the cleaning solvent be hot (180-210°F) and that it circulate at not less than 2 fps. Klein and Atwood[6] found that under these conditions a 5-10 mil layer of Fe_3O_4 dissolved completely in 3 percent ammonium citrate in 6 h. A study of the procedure by Bradley, et al.[42] showed that the rate of dissolution of magnetite was substantially increased by agitation and by the addition of ammonium bifluoride. They also found that in the presence of large amounts of metallic copper dissolution was not always complete unless bromate was used as the oxidizing agent.

The Citrosolv Process is a series of steps in which ferric oxide and magnetite are dissolved at low pH, then metallic copper is brought into solution by an oxidant at high pH. Inhibited citric acid at a concentration of 3-5 percent and an initial pH of about 2.3 is neutralized to a pH of 3.5-4.0 with aqueous ammonia. In this pH range citric acid is essentially present as the monoammonium salt.

---

* Patent assigned to Pfizer Chemicals Div., 235 E. 42nd St., New York, N.Y. 10017.

$$
\begin{array}{l}
\qquad CH_2{-}COO^{-}NH_4^{+} \\
\qquad\ \ | \\
HO{-}C{-}COOH \\
\qquad\ \ | \\
\qquad CH_2{-}COOH
\end{array}
$$

The solution is heated to 180-210°F (the hotter, the better) and circulated at not less than 2 fps for 6–8 h. Next, the circulating solution is cooled to 140-150°F and its pH is raised to 9.5-10.0 by adding aqueous ammonia. When the pH has stabilized in the proper range, 0.5-1.0 percent of sodium nitrite is added to solubilize metallic copper. The solution is then heated to 180-210°F and circulated for 2-4 h, after which the solution is drained and the unit flushed for a few minutes with condensate. Occasionally, triethanolamine is used in place of ammonia, for it also sequesters iron. Some contractors advocate agitating the nitrite solution with air during the copper removal phase.

II.6.b Alkaline EDTA

An excellent method for chemically cleaning boilers with an ammonium salt of EDTA is the subject of several patents(43,44,45) collectively called Alkaline Copper Removal.* This procedure is similar to the Citrosolv Process in that a complexing agent and an oxidizing agent are used to solubilize iron oxide and copper deposits. A commercial solution containing about 40 percent by weight of triammonium EDTA and having a pH of 9.2 is used as the cleaning agent along with a corrosion inhibitor (either sodium 2-mercaptobenzothiazole or sodium gallate) at a concentration of 0.1–0.5 percent.(43) In practice, the boiler to be cleaned is cooled (or heated) until its pressure is about 100 psi. The steam drum is then partially drained to make room for the chelant solution, and the latter, together with the inhibitor, is injected into the boiler through any convenient port. The boiler is fired intermittently for 3–6 h so that its pressure varies from 50 to 150 psi (260-350°F). After cooling to 200°F, the boiler is vented and compressed air is admitted through a lower blowdown or drain line; utility boilers having a capacity of 25,000-50,000 gal of water use about 2000 ft^3 of compressed air admitted over a period of 2-3 h. At the conclusion of the air-blowing, a single rinse with hot (180°F) condensate is sufficient.

Three phases comprise this method for removing boiler scale:

* Patents assigned to Dow Chemical Company, 2020 Dow Center, Midland, MI 48640

Oxide dissolution:

$$3EDTA^{-} + Fe_3O_4 + 4H_2O = Fe(EDTA)^{--} + 2Fe(EDTA)^{-} + 8OH^{-} \quad \text{(II-57)}$$

Air oxidation of ferrous chelonate:

$$2Fe(EDTA)^{--} + \tfrac{1}{2}O_2 + H_2O = 2Fe(EDTA)^{-} + 2OH^{-} \quad \text{(II-58)}$$

Oxidation of metallic copper by ferric chelonate:

$$2Fe(EDTA)^{-} + Cu + EDTA^{-4} = Cu(EDTA)^{--} + 2Fe(EDTA)^{--} \quad \text{(II-59)}$$

Eq. (II-57) shows that 1 lb of iron requires 6.1 lb of $(NH_4)_3EDTA$, or 15.3 lb of a 40-percent solution of the salt. Similarly, Eq. (II-59) indicates that 1 lb of copper consumes 5.4 lb of the salt, or 13.5 lb of a 40-percent solution.

During the course of a cleaning operation concentrations of dissolved iron, residual chelant, pH, and copper are checked frequently by methods given in Chapter XI. The amount of air needed to oxidize ferrous chelonate is found by measuring the oxidation potential of the solution (it should be 175–250 mv) with a saturated calomel electrode and a platinum electrode connected to a digital voltmeter. It is suggested in the original patent that ferrous metal be used as the indicating electrode, but Latimer[(46)] has pointed out that iron electrodes polarize readily, and also, because of random surface strains, different pieces of steel yield large variations in potential measurements.

The foregoing cleaning procedure has a great many advantages over conventional methods of cleaning with acids.

1. The mildly alkaline solution is much safer to handle than acids.
2. Circulation of the cleaning solution is achieved by direct firing of the boiler rather than by an external pump.
3. Deposits are almost completely dissolved, which minimizes subsequent manual clean-up, as there is no precipitation during rinsing.
4. A single rinse with water is sufficient.
5. As the boiler metal is passivated during the air-blowing step, no additional application of a passivating solution is needed.

6. The exhausted solvent can be disposed of by incineration.
7. As a boiler can usually be cleaned in less than 24 h, marine boilers can be cleaned during a routine port call.
8. The corrosion rate of steel in the alkaline solvent is very low—0.3-0.5 mpd.

Hoppe and Parmley[5] point out that although the chemical costs are high for Alkaline Copper Removal, waste disposal costs are considerably lower than they are for some other methods of chemical cleaning.

II.6.c Formaldehyde

The serious hazards of hydrogen sulfide evolved during the cleaning of certain types of equipment with acids were discussed in Chapter I. When the possibility of the evolution of this toxic gas exists, provision must be made to render it harmless, usually by absorption in sodium hydroxide or by flaring. Recently, Frenier, et al.[47] have disclosed a method for rendering hydrogen sulfide innocuous by reacting it with formaldehyde in acid solution, the commercial process being called Dowell Sulfide Scale Removal.* According to the patent, the preferred solvent is inhibited sulfuric acid.

Blatt[48] states that the reaction between formaldehyde and hydrogen sulfide to form sym-trithiane gives a yield of the latter that is 92-94 percent of the theoretical (stoichiometric) amount. This suggests that an alkaline scrubber should be provided for complete safety when using this process for chemical cleaning.

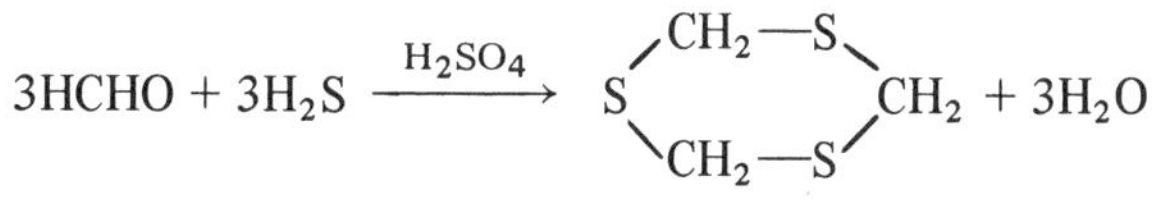

(II-60)

Eq. (II-60) shows that formaldehyde and hydrogen sulfide condense to form a trimer analogous to trioxane. As formaldehyde is highly toxic, producing corneal burns, necrosis of the skin, and if inhaled, pulmonary edema, headache, and gastroenteritis, solutions containing it should be handled with caution.

---

* Patent assigned to Dow Chemical Company, 2020 Dow Center, Midland, MI 48640

II.6.d Ethylenediamine and Hydroxylamine

Cupric and cuprous oxides deposited in steam turbines severely impair the efficiency of the machines. Morris and Call[49] have reported the successful removal of these oxides by flooding turbines with a patented copper solvent[50] containing 18 percent ethylenediamine and 7.5 percent of hydroxylamine hydrosulfate. Both cuprous and cupric ions form stable coordination complexes with ethylenediamine.

$$Cu^{+} + 2NH_2CH_2CH_2NH_2 = Cu[NH_2CH_2CH_2NH_2]^{+} \qquad K = 6.3 \times 10^{10} \qquad \text{(II-61)}$$

$$Cu^{++} + 2NH_2CH_2CH_2NH_2 = Cu[NH_2CH_2CH_2NH_2]^{++} \qquad K = 1.0 \times 10^{20} \qquad \text{(II-62)}$$

The role of hydroxylamine in the solvent is uncertain, but it readily decomposes to ammonium ion and nitrous oxide; as indicated by Eq. (II-63), the reaction is favored in alkaline solution.

$$4NH_3OH^{+} = N_2O + 2NH_4^{+} + 3H_2O + 2H^{+} \qquad \text{(II-63)}$$

$$4NH_2OH = N_2O + 2NH_3 + 3H_2O \qquad \text{(II-64)}$$

References

(1) Rabald, E. 1968. *Corrosion guide.* New York: Elsevier Publishing Co.

(2) Hantelman, W. D. 1972. Ten years experience in pre-operational chemical cleaning of power plant steam generating systems. *Proc. Int'l. Water Conf., Eng. Soc. W. Pa.* 33:101.

(3) Atwood, K. L. 1972. A review of pre-operational cleaning practices in high pressure utility systems. *Proc. Int'l. Water Conf., Eng. Soc. W. Pa.* 33:73.

(4) Duncan, F. R., Fyock, D. H., Baldwin, L. V., and Hopkins, R. D. 1972. Preoperational cleaning of high pressure boilers of General Public Utilities Corporation. *Proc. Int'l. Water Conf., Eng. Soc. W. Pa.* 33:97.

(5) Hoppe, T. C. and Parmley, J. G. 1972. Preoperational chemical cleaning of utility systems operating at 2000 psi and higher. *Proc. Int'l. Water Conf., Eng. Soc. W. Pa.* 33:63.

(6) Klein, H. A. and Atwood, K. L. 1964. Chemical cleaning of utility boilers. *Proc. Am. Power Conf.* 26:762.
(7) Engle, J. P., Floyd, G. L., and Rosene, R. B. Feb., 1959. Chloride stress corrosion cracking of austenitic stainless steels. *NACE Publication 8A259, Corrosion* 15:69t.
(8) Uhlig, H. H. 1967. *Corrosion and corrosion control.* New York and London: John Wiley & Sons, Inc.
(9) Dean, J. A. *Lange's handbook of chemistry.* 1974. 11th ed. New York: McGraw-Hill Book Co.
(10) Gruse, W. A. and Stevens, D. R. 1960. *Chemical technology of petroleum.* 3rd ed. New York: McGraw-Hill Book Co.
(11) Axsom, J. F. Mar. 14–18, 1977. Energy savings via semi on stream cleaning of crude preheat exchangers. Paper No. 70. *Int'l. Corr. Forum.* Houston, Texas.
(12) Maloney, J. M. 1954. Internal cleaning of boilers. *Ind. Eng. Chem.* 46:983.
(13) Yorkgitis, E. A. 1980. Fluoride cleaning of steel-plant cooling-water systems. *Proc. Int'l. Water Conf., Eng. Soc. W. Pa.* 41.
(14) McCoy, J. W. 1974. *The chemical treatment of cooling water.* New York: Chemical Publishing Co.
(15) Hoar, T. and Holliday, R. 1953. The inhibition of quinolines and thioureas of the acid dissolution of mild steel. *J. Applied Chem.* 3:502.
(16) Trabanelli, G. and Carassiti, V. 1970. Mechanism and phenomenology of organic inhibitors, in *Advances in corrosion science and technology.* Vol. 1, Fontana, M. G. and Staehle, R. W., eds. New York–London: Plenum Press.
(17) Branch, G. E. K. and Calvin, M. 1944. *The theory of organic chemistry.* New York: Prentice-Hall.
(18) Riggs, O. L., Jr. 1973. Theoretical aspects of corrosion inhibitors and inhibition, in *Corrosion inhibitors.* Nathan, C. C., ed. Houston: National Association of Corrosion Engineers.
(19) Mayanna, S. M. and Setty, T. H. V. 1975. Effect of benzotriazole on the dissolution of copper single crystal planes in dilute sulfuric acid. *Corros. Sci.* 15:627.
(20) Monroe, R. F., Kucera, C. H., Oakes, B. D., and Johnson, N. G. July 14, 1960. *Compositions for inhibiting corrosion.* U.S. Patent No. 3,077,454.
(21) Teumac, F. N. Dec. 3, 1968. *Acid corrosion inhibitors (thiomorpholine and phenothiazine and derivatives).* U.S. Patent No. 3,414,521.
(23) Hackerman, N., Hurd, R. M., and Annand, R. R. Jan., 1962. Some structural effects of organic N-containing compounds

on corrosion inhibition. *Corrosion* 18:37t.

(24) Crouse, D. N. Oct. 8, 1979. Danger with sulfuric acid (letter). *Chem. Eng. News* 57 (41): 4.

(25) Kucera, C. H. Apr. 14, 1970. *Corrosion inhibitor for aqueous hydrochloric acid solutions.* U.S. Patent No. 3,506,581.

(26) Mannich, V. C. and Krösche, W. 1912. Ueber ein Kondensationsprodukt aus formaldehyd, ammoniak und antipyrin. *Archiv der Pharmazie* 250:647.

(27) Vosta, J., Pelikun, J., and Smrz, M. 1974. Investigation of corrosion inhibitors for iron in hydrofluoric acid. *Werkst. Korros.* 25 (10):750.

(28) Mansfield, F. and Kenkel, J. V. 1975. Organic corrosion inhibitors for titanium. *Corr. Sci.* 15:767.

(29) Report NACE T-8A. May, 1962. Evaluation of inhibitors used to prevent corrosion of metals by acid cleaning methods. *NACE Publication 8A 162. Materials Protection* 1:107.

(30) Steinhauer. A. F. Sept. 30, 1958. *Method of making alkyl diphenyl ether sulfonates.* U.S. Patent No. 2,854,477.

(31) Feighner, G. C. 1976. Alkylsulfonates, in *Anionic surfactants.* Vol. 7, Part 1, W. M. Linfield, ed. New York: Marcel Dekker, Inc.

(32) Poff, G. Oct., 1978. The effect of ammonium bifluoride on the corrosion rates of some common chemical cleaning solvents. *Materials Performance* 17 (10):24.

(33) Mayer, L. and Guelich, J. 1963. Hydrogen fluoride (HF) inhalation and burns. *Arch. Environ. Health* 7:445.

(34) Martin, R. C. and Abel, W. T. Nov. 8, 1960. *Copper and iron containing scale removal from ferrous metal.* U.S. Patent No. 2,959,555.

(35) Engle, J. P. and Fitzgerald, R. W. Aug., 1959. Single-stage copper removal treatment. *Power Eng.* 63 (8):76.

(36) Cain, C., Jr. and Schonhoff, C. C. 1964. Chemical cleaning experience at steam-electric power plants of the Tennessee Valley Authority. *Proc. Am. Power Conf.* 26:779.

(37) Wackenhuth, E. C. and Richards, H. E. 1966. Recent experiences in chemically cleaning a modern boiler. *Proc. Am. Power Conf.* 28:785.

(38) Alfano, S. Jan. 8, 1963. *Process for removing copper-containing iron oxide scale from metal surfaces.* U.S. Patent No. 3,072,502.

(39) Alfano, S. and Bell, W. E. 1961. Chemical removal of magnetite and copper. *Proc. Int'l. Water Conf., Eng. Soc. W. Pa.* 22.

(40) Alfano, S. and Bell, W. E. Dec., 1961. Faster chemical clean-

ing for high pressure boilers. *Power Eng.* 65 (12):66.

(41) Alfano, S. Nov., 1962. Cleaning the iron system using ammoniated citric acid. *Combustion* 33 (11):17.

(42) Bradley, G. W., Martin, L. D., Banks, W. P., and Frost, J. G. 1975. Investigation of ammonium citrate cleaning solvents. *Proc. Int'l. Water Conf., Eng. Soc. W. Pa.* 36:183.

(43) Lesinski, C. A. Mar. 7, 1967. *Scale removal, ferrous metal passivation and compositions therefor.* U.S. Patent No. 3,308,065.

(44) Teumac, F. N. Nov. 26, 1968. *Passivation of ferrous metal surface.* U.S. Patent No. 3,413,160.

(45) Harriman, L.W. and Muehlberg, P. E. Apr. 15, 1969. *Removal of copper containing incrustations from ferrous surfaces.* U.S. Patent No. 3,438,811.

(46) Latimer, W. M. 1952. *The oxidation states of the elements and their potentials in aqueous solutions.* 2nd ed. Englewood Cliffs: Prentice-Hall.

(47) Frenier, W. W., Coffey, M. D., Huffines, J. D., and Smith, D. C. Sept. 2, 1980. *Composition and method for removing sulfide-containing scale from metal surfaces.* U.S. Patent No. 4,220,550.

(48) Blatt, A. H., ed. 1943. *Organic synthesis,* collective Vol. 2, pp. 610–611. New York: John Wiley & Sons.

(49) Morris, E. B. and Call, R. G. Chemical cleaning of turbines for removal of copper oxide deposits. *ASME Paper 64-WH/BFS-1.*

(50) Call, R. G. Oct. 10, 1961. *Cleaning composition and a method of its use.* U.S. Patent No. 3,003,970.

III

Nonaqueous Cleaning Solvents

It is exceptional in chemical cleaning that inorganic fouling is completely dissolved by any single aqueous solution. By properly selecting a non-aqueous solvent, however, it is often possible to dissolve organic material almost completely obtaining, in effect, a true solution. Nonaqueous solvents are circulated the same as aqueous solutions except that hoses and rubber-lined tanks cannot be used, as these materials are softened and destroyed by hot organic solvents; for this reason, the chemical cleaning truck must be piped to the unit to be cleaned, rather than being hose-connected. When cleaning tubes in certain exchangers in petroleum refineries it is sometimes possible to use an existing hot oil system on the shell side of the exchanger to heat the solvent; in other applications heat is supplied through a small heat exchanger. In either case, the recirculating tank is purged and blanketed with nitrogen while being vented to a hose coiled in a barrel of cold water to condense any vapors evolved from the tank. When cleaning with flammable solvents, the recirculating tank must be grounded to prevent any accumulation of static electrical charge.

Unlike acidic and oxidizing solutions, organic solvents dissolve organic materials without decomposing them and, thus, are especially useful for removing mineral and vegetable oils, waxes, asphalt, asphaltenes, pitches, tars, bitumens, and polymerized species from chemical and petroleum processing units. Continuous circulation is essential when using organic solvents as the fill-and-soak procedure is very inefficient. Naphthas and aromatic mixtures are useful for cleaning heat exchangers fouled with organic matter, but for columns and reactors vapor phase degreasing, using chlorinated alkanes, is more appropriate. For lack of a better place, acid foaming is briefly described in Section III.2; safety aspects of nonaqueous chemical cleaning are considered in Section III.3.

III.1 TYPES OF SOLVENTS

In the interest of safety, flammable solvents having flash points(1) less than 150°F should not be used for cleaning in the open air. Although this restriction does not apply to closed systems that are suitably piped and vented, it is always best to use solvents having high flash points for chemical cleaning because of the difficulty of safely handling the more volatile solvents. For practical purposes, it is usually possible to use one of three types of organic solvents for industrial chemical cleaning. These are kerosene naphtha, boiling range 430-610°F, flash point 185°F; heavy grade aromatic solvent, boiling range 540-765°F, flash point 280°F; chlorinated hydrocarbons, which are not flammable.

III.1.a Naphthas

Paraffins of high molecular weight, such as waxes, dissolve reasonably well in hot naphtha circulated for 3-4 h, preferably at a temperature above the melting point of the wax. Propylene polymers, polybutenes, and similar materials also dissolve well in this type of solvent. The use of naphthas is particularly advantageous in petroleum refineries because the used solvent can be transferred to recovered oil or crude feed tankage for reprocessing, thus avoiding a heavy disposal expense. It is beneficial to add 0.1 percent each of an oil-soluble and a water-soluble surfactant to the solvent before use, for this combination facilitates water rinsing after the nonaqueous solution has been pumped out.

III.1.b Aromatic Solvents

For dissolving residual material, a heavy grade of aromatic petroleum solvent is appropriate. This type of solvent contains 80-90 percent of C_9^+ aromatics, that is, compounds that contain nine or more carbon atoms including six in a benzene ring. Table III-1 lists several of these compounds, the most plentiful being the three trimethylbenzene isomers.

TABLE III-1

Heavy Aromatic Hydrocarbons

Name	*Carbon no.*	*Boiling point* °C	°F
1,2,3-Trimethylbenzene	9	176	359
1,2,4-Trimethylbenzene	9	169	338
1,3,5-Trimethylbenzene	9	165	328
o-Methylethylbenzene	9	165	330
n-Propylbenzene	9	159	318
o-Diethylbenzene	10	183	362
n-Butylbenzene	10	183	362
n-Amylbenzene	11	205	402
o-Diisopropylbenzene	12	210	410

Mixtures of these relatively high-boiling aromatics are useful for dissolving heavy, viscous materials that themselves contain condensed aromatic rings, including asphalts, asphaltenes, pitches, tars, and the like. Higher boiling solvents containing approximately 60 percent aromatics and 40 percent paraffins and naphthenes are available that sometimes perform well with mixed organic fouling; this type of solvent was once called high aromatic naphtha.[(2)] As the approximate boiling range of aromatic naphtha is 540–765°F, it can be circulated in a closed system at about 275°F without danger of flashing, which is a considerable advantage in bringing heavy organics into solution. As with naphthas, the used aromatic solvents can be transferred to tankage for reprocessing in a petroleum refinery, thus solving what can be a major disposal problem.

III.1.c Chlorinated Hydrocarbons

Chlorinated methanes and ethanes are especially useful for removing organic material from tray columns, reactors, and other vertical vessels. This is done by introducing a suitable vaporized solvent into the top of the vessel, where it condenses, then flows down over the internal surfaces and collects as a liquid in the bottom. The used solvent is next pumped into a small steam-heated still where it is rectified and the vapor is recycled to the top of the vessel being cleaned, a continuous process analogous to a Soxhlet extraction in the laboratory. Eventually, the organic fouling is concentrated in the still bottoms and the reclaimed solvent is collected in the bottom of the cleaned vessel. This is an economical procedure in that not much of the expensive solvent is used and that virtually all of it is reclaimed, and can be reused. The organic material can be washed out of the still with naphtha or aromatic solvent to be incinerated or reprocessed, as appropriate. On account of the possibility of generating phosgene, $COCl_2$, an extremely toxic gas, care must be taken to remove as much as possible of the chlorinated solvent from the organic residue before it is incinerated. Unless arrangements can be made for reclaiming chlorinated solvents, it is almost prohibitively expensive to use them because of their high initial cost and the large fees charged to dispose of these waste solutions at an approved hazardous waste disposal site. Also, it should be mentioned that chlorinated solvents demand better pump seals than those normally installed for use in acid cleaning, and teflon, rather than neoprene, hoses must be used.

All chlorinated solvents are irritating to the eyes and mucosa, some exert a narcotic effect at high concentrations, and others can damage the liver and kidneys. Table III-2 lists names, formulas, and properties of several chlorinated compounds that have been used for vapor-phase degreasing.

TABLE III-2

Properties of Chlorinated Hydrocarbons

Name	*Formula*	*Threshold limit value**	*Boiling point* °C	°F
Methylene chloride	CH_2Cl_2	250	40	104
Chloroform	$CHCl_3$	10	62	143
Carbon tetrachloride	CCl_4	10	77	170
1,1,1-Trichloroethane	CH_3CCl_3	350	74	165
1,1,2-Trichloroethane	CH_2ClCCl_2	10	114	236
Trichloroethylene	$CHCl{=}CCl_2$	100	87	188
Tetrachloroethylene	$Cl_2C{=}CCl_2$	100	121	250

* Maximum ppm in air for chronic exposure (8 h/day, 40 h/week).

Moulton[3] reports that trichloroethylene is more effective than 1,1,1-trichloroethane for removing high-melting pitches and waxes, although it is also more of a health hazard. Dale[4] has also published some useful data on nonflammable solvents. The Threshold Limit Values in Table III-2 are those quoted by Plunkett,[5] but it should be realized that these are merely guides and that unusual sensitivities may render some individuals much less tolerant than the average. Nevertheless, it seems reasonable to make use of those solvents having the higher Threshold Limit Values, namely, methylene chloride and 1,1,1-trichloroethane. Although the former is the better all-around solvent, its boiling point is too low for use in warm weather so, in general, 1,1,1-trichloroethane is suggested for use in warm weather while methylene chloride should be the choice in cool or cold weather.

Potential users of chlorinated solvents should be aware of their corrosive effect on certain metals. These solvents chemically attack the more electropositive metals, such as aluminum and zinc, at defects in the normally protective oxide films on the metal surfaces, yielding the anhydrous metal chloride and a dimer derived from the chlorinated compound. Thus, the reaction between aluminum and carbon tetrachloride can be written as

$$2Al + 6CCl_4 \longrightarrow 3C_2Cl_6 + 2AlCl_3 \qquad \text{(III-1)}$$

A few ppm of water in the solvent greatly speeds the reaction, which in some instances can be explosive. Zinc, too, is completely removed from brass by boiling carbon tetrachloride containing a few ppm of water.

Many chemicals, among them malononitrile, dioxane, pyridine aldehydes, and alkyldiones inhibit the reaction in Eq. (III-1) by forming chemisorbed films on the metal surface, thus protecting the metal during vapor phase degreasing. Care must be taken in selecting mixed solvents, however, to ensure that the inhibiting film is not soluble in the solvent, otherwise there can be no inhibition and aluminum or zinc corrode very rapidly. Archer[6] lists some 60 corrosion inhibitors used with chlorinated solvents; the more common of these solvents can be purchased already suitably inhibited.

III.2 ACIDIC FOAMS

Chemical cleaning with acidic foams is a proprietary procedure* that has several advantages, particularly in cleaning large vessels such as surface condensers and boilers. This method has been used for a number of years in

* Dow Chemical Company, 2020 Dow Center, Midland, Michigan 48640

oil-field applications, especially the removal of deposits from boreholes[7] and also to some extent for etching ferrous metals. Balezin, et al.[8] report that ferric ion is transferred almost completely to the foam layer during pickling of oxidized steel. Foam cleaning of a boiler requires only about one-third of the solvent that the fill-and-soak or fill-and-circulate procedures use, and very large condensers can be cleaned in 8–12 h by foaming, whereas 3–5 days may be required for conventional methods.

Careful metering of water, nitrogen, foaming agent, and acid is necessary to avoid excessive foaming and to ensure a wet foam rather than one of heavy consistency resembling shaving cream. In practice, one volume of acid solution is expanded to about ten volumes of foam, which is passed through the unit to be cleaned, covering the internal structure with acidic solvent, and rapidly dissolving deposits. The selection of a suitable foaming agent is critical to the success of acidic foaming; it must produce a foam that is stable for a reasonable length of time, i.e., 5–10 min, but that can be readily destroyed for disposal. Amides, prepared by reacting fatty acids with amines, are used as foam boosters in some applications. Dihydroxyethyllauramide, for example, formed by reacting lauric acid with diethanolamine, is a very weakly basic compound that is used as a foam stabilizer for some purposes.

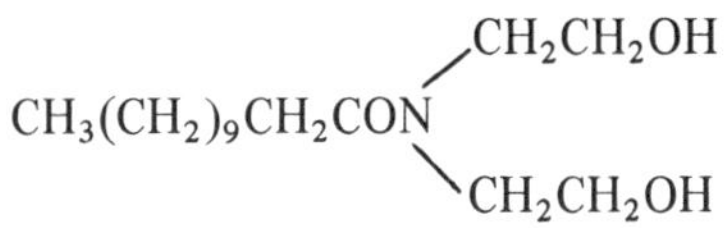

Dihydroxyethyllauramide

Foam Frac,* which is used in the proprietary procedure, has an isoelectric point in the range of pH 1.7–3.5, where it is at minimum solubility, so it tends to precipitate during rinsing. The temperature limit for acidic foaming is about 160°F, and the success of this method of cleaning depends to a great extent on the use of a patented device[9] for metering and mixing the ingredients.

III.3 HAZARDS OF NONAQUEOUS SOLVENTS

As remarked earlier, safety considerations preclude the use in open air of solvents having a flash point less than 150°F. The flash point is the temperature at which the vapor above a heated portion of a flammable sol-

* Dow Chemical Company, 2020 Dow Center, Midland, Michigan 48640

TABLE III-3

Systemic Effects of Chlorinated Hydrocarbons

Name	*TLV**	*Anemia*	*Coma*	*Cardiac depression*	*Cardiac arrest*	*Ventricular fibrillation*	*Hepatic necrosis*	*Renal demage*	*Carcinogen*
Methylene chloride	250	+	–	–	–	–	–	–	–
Chloroform	10	–	–	+	–	–	+	+	+
Carbon tetrachloride	10	–	+	+	–	+	+	+	–
1,1,1-Trichloroethane	350	–	–	–	–	–	+	+	–
1,1,2-Trichloroethane	10	–	–	–	–	–	+	+	–
Trichloroethylene	100	–	–	–	+	+	–	–	+
Tetrachloroethylene	100	–	+	+	–	–	+	+	–

* Threshold Limit Value, ppm in air

vent momentarily ignites when a small pilot flame is passed across the surface. The higher its flash point the safer a solvent is to use. When cleaning with naphthas and aromatic solvents heat and open flames must be kept away, smoking is prohibited, and the chemical cleaning truck and the unit being cleaned must be electrically bonded and grounded to prevent static electrical discharge. As all organic solvents are to some extent toxic, they should be used with adequate ventilation, and workmen should be cautioned to avoid prolonged or repeated breathing of vapors or contact of liquid with the skin. Solvents are especially dangerous when inhaled, for in addition to inflicting local damage on the lungs and respiratory mucosa, they rapidly pass into the circulatory system exerting a depressing effect on the central nervous system manifested by lassitude, fatigue, stupor, unconsciousness, and in some cases, coma.[(5)]

In addition to depressing the central nervous system, all of the chlorinated hydrocarbons cause eye and respiratory irritation, dermatitis resulting from defatting, nausea, vomiting, dizziness, and headache. Along with these unpleasant general symptoms, individual compounds produce more dangerous systemic effects, the more important of which are indicated in Table III-3.

Trichloroethylene is particularly hazardous for in the presence of air and alkali it forms phosgene and dichloroacetylene,[(5)] both of which are highly toxic.

$$2Cl_2C{=}CHCl + 3/2O_2 \longrightarrow 2COCl_2 + ClC{=}CCl + H_2O \quad \text{(III-2)}$$

In addition to the specific toxic effects mentioned, allergic reactions to chlorinated hydrocarbons are common with individual susceptibility subject to wide variations. As these solvents readily pass through the skin, gloves and protective clothing as well as respiratory protection are mandatory.

References

(1) Am. Soc. Testing Materials. 1980. Standard Method of test for flash point by Pensky-Martens closed tester. *Annual Book of ASTM Standards.* Pt. 17. D93.

(2) Durrans, T.H. 1971. *Solvents.* 8th ed. Revised by E. H. Davies. London: Chapman and Hall Ltd.

(3) Moulton, F. H. Feb., 1979. Vapor degreasing with chlorinated solvents. *Product Fin.* 32 (2):16.

(4) Dale, F. H. Feb., 1979. Solvent cleaning. *Product Fin.* 32 (2): 19.

(5) Plunkett, E. R. 1976. *Handbook of industrial toxicology.* New York: Chemical Publishing Co.

(6) Archer, W. L. Feb., 1979. Comparison of chlorinated solvent-aluminum reaction inhibitors. *Ind. Eng. Chem. Prod. Res. Dev.* 18 (2):131.

(7) Sabirov, Kh. Sh., Ilyukov, V. A., Ugolev, V. S., Pastukov, I. V., and Ageev, L. S. 1973. Increasing the distance a deposit is penetrated by acid-foam treatment of boreholes. *Nefteprom. Delo.* 11:11 (USSR) (*CA* 81:27832g).

(8) Balezin, S. A., Bol'shakova, G. S., and Gorin, L. F. 1970. Foaming agents of etching solutions as regulators of the concentration of iron (3+) ions during the etching of ferrous metals. *Nauch. Tr. Novosibirsk. Gos. Pedagog. Inst.* 54:27 (USSR) (*CA* 79:117336m).

(9) Simmons, B. G. Jan. 9, 1979. *Apparatus for making foamed cleaning solutions and methods of operation.* U.S. Patent No. 4,133,773.

IV

Nonchemical Cleaning Procedures

Although it may appear from the preceding chapters that chemical cleaning with acids or other solvents is the major method of industrial cleaning, a large volume of such work is still done using mechanical methods or jetting with high-pressure water. It is a more-or-less standard procedure, for instance, to wire-brush boiler drums, followed by water jetting to permit inspection for pitting and cracking.

IV.1 MECHANICAL METHODS

Tubes of drum boilers have for many years been cleared of scale by turbining, a procedure in which rotating cutters are pushed through the tubes, crushing and cutting through scale. The cutters are driven by small turbines powered by steam, water, or preferably, compressed air supplied through a hose attached to the motor. The latter must be lubricated and cooled continuously while being operated and it must be kept constantly in motion to avoid damaging tubes. Universal joints permit the cleaning of curved tubes in smaller steam generators, but mechanical cleaning of large utility boilers is not feasible because of their size. The expenditure of time and labor is prohibitive in any case and the small diameter and sharp bends of the tubes of modern boilers demand chemical cleaning. Water tubes are turbined from the drums, after which the lower portions of the boiler must be flushed with water to remove broken scale from the mud drum, lower water wall headers, and blowdown lines before start-up.

Knocker heads, also driven by small motors, are appropriate for removing scale from fire-tube boilers.[1] The air-driven motor vibrates the head violently against the inside of the tube, loosening scale on the outside (the water side); hard, brittle, water-formed scales that are friable respond well to knocker heads.

In most instances, furnace tubes are cleared of slag by mechanical chipping. Lee, et al.,[(2)] however, found that furnace slag is bound to superheater and other tubes by water-soluble sulfates. Slag itself contains a great number of small interconnecting pores, which although impervious to water alone, are permeated by water containing 0.1-0.2 percent of a nonionic wetting agent. By saturating the slag in a furnace with water containing a wetting agent, and allowing it to stand for a day or two, part of the binding salt is dissolved, and the bulk of the slag becomes susceptible to removal by high-pressure water jetting. This simple procedure, where applicable, saves considerable labor.

Sandblasting is sometimes useful for removing scale, grease, and other debris from piping, and slag from furnace tubes. Nunciato, et al.[(3)] have disclosed a procedure for decoking furnace tubes by injecting steel shot entrained in a stream of nitrogen moving at the high velocity of 230-330 fps. When sand is substituted for steel shot the method is called the SANDJET* cleaning process. This proprietary method is especially effective for cleaning piping runs. One end of the pipe, or some other convenient entry point, is fitted with a 150-lb flange to permit connecting to the abrasive injecting equipment. The line should be substantially dry, with all lateral connections blinded off, and all valves replaced with spools. A flow of water is necessary at the discharge end of the pipe to suppress dust, and some sort of a refuse container, e.g., a dumpster, is needed to receive the abrasive and cleaning products. The flow of nitrogen propellant is established and the abrasive is injected into the propellant stream for approximately 5 min, which is often sufficient time to produce a clean pipe. The advantage of this method is that all material connected with the operation is inert and, thus, not difficult to dispose of.

IV.2 HIGH-PRESSURE WATER JETTING

In many instances tube bundles in heat exchangers can be cleaned faster, at lower cost, and more safely using a jet of high-pressure water than by chemical cleaning.[(4)] The internal surfaces of plugged tubes can usually be cleared by inserting a ¼-in. pipe lance fitted with a special nozzle that

---

* Union Carbide Industrial Services Company, 308 Harper Drive, Moorestown, New Jersey 08057.

sprays water straight ahead to remove plugging and also around the periphery of the nozzle to clean the walls of the tube. When water is delivered to the orifices at up to 10,000 psi, the impact and erosive force of the jets of water clears scale, deposits, and other obstructions from the tube. Bubbles of water vapor and air, formed during the 10,000-psi pressure drop as the jet emerges from the tip, implode on the surface being cleaned causing some cavitation damage. If a small area of a steel surface is subjected to a continuous jet of water normal to the surface, metal is eroded at about 0.4 mil per second. For this reason, it is important to keep the jet moving when water jetting industrial equipment.

As pipe lances must be pushed through plugged tubes, it is necessary to provide staging for cleaning tube bundles above ground level so that good footing is available to the operator. If it is not convenient or practicable to stage, tubes can be cleaned without pulling the bundle out of its shell by using flexible lances fitted with a tip having several openings of about 0.024 in. each. In a typical configuration one orifice is directed forward and two others are directed at 45° angles from the forward orifice to assist in propelling the lance through the tube.

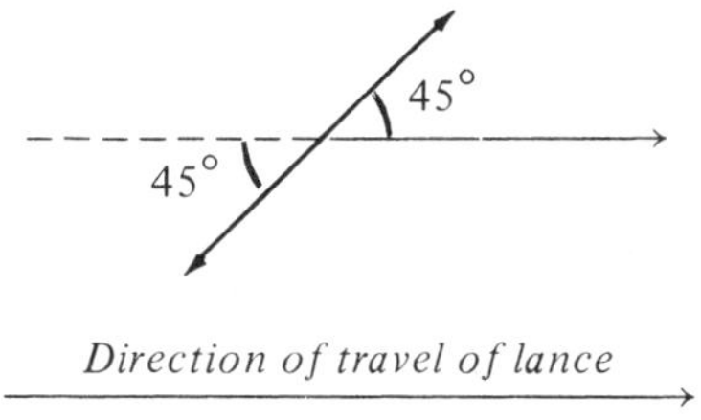

These nozzles direct about 40 percent of the water forward and about 60 percent rearward. This results in a net forward impulse of about 40 lb, which first pushes the lance into a tube, then exerts a strong scouring effect when it is pulled out of the tube. Flexible lances, which are 3/8-in. hoses bound with stainless steel mesh, are limited to 7000 psi when in constant use because the continual flexing, pulling, and pushing weakens them rapidly; they are perhaps ten times more expensive than pipe lances. As flexible lances can be operated from a very small platform, for instance, the limited staging required for removing channel covers, a great deal of labor is saved by their use. The usual rate of water usage is 12-13 gpm, and 80-120 tubes per hour can be cleared, depending upon the condition of the bundle.

The outside of a tube bundle can be "shotgunned" by hand using a gun-like apparatus equipped with a dead man trigger that, when released, automatically stops the pump, and a dump valve to eliminate instantly all flow to the nozzle. The jet tip is a straight steel nozzle designed to produce a narrow, collimated stream; a fan tip can be substituted, if desired, for a softer, broom-like action. When a number of bundles are to be cleaned on the shell side, it is advantageous to use a mechanical bundle roller, which rolls the bundle slowly while a mechanical drive moves a compound spray head over the bundle horizontally from end to end at the rate of 1 fps. It should be mentioned that triangularly pitched tube bundles cannot be cleaned by high-pressure water jetting because the interior tubes are inaccessible to the jets of water.

Pipelines from 2 to 8 in. in diameter can be cleaned using ordinary high-pressure connecting hose attached to a come-along nozzle or mole, which can pass through one 90° bend or two 45° bends. For larger pipe a larger adjustable nozzle can be used with a greater volume of water obtained by using two pumps.

Water jetting can also be used for a number of purposes not usually associated with chemical cleaning. It has been used to remove coke, tar, asphalt, fiberglass, and polybutene, as well as to excavate earth from beneath pipelines to prevent corrosion induced by the wet soil. With some materials a number of small orifices (0.026 in.) is more efficient than a single large jet, but with others the reverse is true, so it is advisable to experiment with nozzles and pressures for each new application. In general, no more pressure should be used than that necessary to shear the fouling material from the substrate. Excessive pressure wastes energy, wears out equipment, and makes a hand-held jet more difficult and dangerous to handle. Tursi and DeLecce[5] suggest the following operating conditions for effective high-pressure water jetting of boiler tubes.

1. Use flexible lances ¼-in. in diameter and 14–20 ft long.
2. Nozzles should have at least 10 orifices evenly spaced around the circumference.
3. Water flow rate should be at least 10 gpm at 7000 psi pump pressure.
4. The pump drive motor should deliver a minimum of 75 horsepower.
5. Use supply hoses having a minimum bursting pressure of 30,000 psi.

When removing oily or greasy materials the addition of a few parts per million of detergent to the water facilitates cleaning. When water jetting boiler tubes to remove the debris inevitably remaining after chemical cleaning, the tubes rust very rapidly, even if previously passivated, as the jet quickly strips the passive film thus exposing the steel surface to oxidation. Tursi and DeLecce[(5)] found that adding 1 lb of sodium nitrite per 100 gal of water used for jetting boiler tubes eliminated rusting completely.

Several years ago Kudin, et al.[(6)] reported that after adding polyacrylamide (MW = 2×10^6) to water at a concentration of 1000 ppm, the resulting solution was capable of rapidly perforating a steel plate 2.5 mm thick when directed against it in a jet at 250 atmospheres (3675 psi). Their explanation of this phenomenon is that aggregates of polymer and water molecules, when moving rapidly, behave as undeformable particles. That is, when moving relatively slowly the aggregates behave as droplets of liquid, but when moving rapidly, as in a high-pressure jet, they behave as solid particles. This suggests that the addition of a polymer to water used for water jetting might improve the cutting capability of the stream. Polymer does, indeed, do this, improving the effectiveness of the procedure up to 50-fold and, in addition, focusing the jet so that before diverging it can reach fouled tubes deep within a square-pitched tube bundle. Polymer also reduces friction in the supply line and nozzle of the lance and–unexpectedly–reduces erosion of the metal being cleaned.[(7)] The presence of polymer increases the length of time that a metallic surface can be exposed to a jet of high-pressure water by a factor of nine over water without polymer. This increase in the induction period of erosion is significant in that during water jetting the jet traverses the surface being cleaned at approximately 1 fps and the residence time of the water jet at any point on the surface is much less than the induction period. The increased induction period and the decreased initial erosion rate obtained by adding polymer protect the metal surface from overly long exposure to the high-pressure jet, minimizing removal of surface metal and permitting safer, more thorough cleaning.

Dr. W. G. Howells* speculates that the collimation of the jet and the reduction of friction attributable to the presence of polymer are the result of induced laminar flow both before and after the jet of water emerges from the nozzle. In his view, the laminae consist of linearly aligned polymer molecules surrounded by extended clusters of hydrogen-bonded water

* Berkeley Chemical Research Inc., P.O. Box 9264, Berkeley, CA 94709. Private communication.

molecules. As these molecules strike a surface at high speed they are essentially rigid and, thus, intensify the impact of the jet manyfold.

Best results are realized by using a water-in-oil emulsion of an anionic polyacrylate-polyacrylamide copolymer having a molecular weight of about six million at a concentration of 3000 ppm. To obtain precise mixing without degrading the macromolecular structure, the polymer must be proportionally metered through a corrosion-resistant gear pump into the water supply, which then passes through a static mixer into a supply tank mounted on the water pump. Passage through the static mixer induces a phase inversion resulting in an oil-in-water emulsion; it is imperative that the polymer be metered into the water as reverse addition causes instantaneous solidification of the emulsion. The cost of this essential metering and injection equipment is currently 1500–3500 dollars. The mixer and polymer pump should be flushed with kerosene immediately after use to remove any residual polymer emulsion, which would otherwise solidify on standing, rendering the equipment inoperable. Flushing with water or any alcohol-containing solvent is inappropriate for removing the concentrated polymer emulsion for it is thereby converted to a solid mass.

When carrying out high-pressure water jetting it is necessary to be aware that it is a hazardous procedure, that all safety precautions must be taken to ensure that proper equipment including scaffolding is supplied, and that the area is roped off to prevent the entrance of anyone unaware of the hazardous operation. At a pressure of 9000 psi and a flow rate of 12 gpm through a 0.067-in. nozzle, a jet of solution is produced that moves at 1100 fps. Workmen doing the jetting should not be hurried; attempting to clean tubes as quickly as possible is eminently dangerous and invariably yields unsatisfactory cleaning. Lances must be passed in and out deliberately, especially when withdrawing the lance from a tube. In the interest of safety, rigid lances only should be used for polymer solutions.

References

(1) Woodruff, E. B. and Lammers, H. B. 1967. *Steam plant operation.* 3rd ed. New York: McGraw-Hill Book Co.

(2) Lee, G. K., Mitchell, E. R., Grimsey, R. G., and Hopkins, S. E. 1964. An investigation of fuel-oil additives to prevent superheater slagging in Naval boilers. *Proc. Am. Power Conf.* 26:531.

(3) Nunciato, D. J., White, N. H., and Woodburn, W. A. May 20, 1980. *Method for decoking fired heater tubes.* U.S. Patent No. 4,203,778.
(4) Goodwin, R. J. Aug. 2, 1976. High pressure water speeds refinery cleanups, cuts turn around costs. *Oil and Gas J.* 74 (31): 134.
(5) Tursi, T. P., Jr. and DeLecce, R. J., Jr. 1976. Development of very high pressure water jet for cleaning Naval boiler tubes. *Marine Technology* 13 (3):272.
(6) Kudin, A. M., Barenblatt, G. I., Kalashnikov, V. N., Vlasov, S. A., and Belokon, V. S. Oct. 8, 1973. Destruction of metallic obstacles by a jet of dilute polymer solution. *Nature* 245:95.
(7) Sedahmed, G. H., Abd El-Naby, B. A., and Abdei-Khali, A. 1977. Corrosion inhibition through decreasing the rate of mass transfer by drag-reducing agents. *Corr. Sci.* 17:865.

V

Passivation

After being cleaned with acids, metal surfaces, especially those of steel, are in a highly active state in which they are subject to rapid oxidation in the presence of air. This can be prevented by impressing a positive electrical potential on the metal or by treating it with certain chemicals, either of which procedures induces the formation of a protective, nonconducting film of iron oxide on the surfaces of the acid-cleaned steel equipment. As the passivation potentials of chromium, nickel, copper, brass, monel, and cupronickel are much lower than that of iron, these metals are essentially self-passivating, spontaneously forming an oxide film when exposed to air. Metals covered or partially covered by oils, scales, deposits, or copper foils cannot be passivated so it is most important that new and retubed boilers and heat exchangers be thoroughly cleaned before being passivated and put into service; for best results passivating should be done immediately before the equipment is to be used. Chemically cleaned or water-jetted boilers and heat exchangers should always be passivated to avoid excessive fouling by iron oxides in service.

V.1 THEORETICAL CONSIDERATIONS

Passivity of iron surfaces is brought about by any of several strong oxidizing agents, including nitrite, chromate, molybdate, and tungstate, all of which are capable of oxidizing ferrous ion, thus permitting metallic iron to dissolve only in the ferric state. This gives rise to a stable protective film of oxide that, being semiconductive, inhibits further dissolution. In the presence of phosphates, oxygen fulfills the same function, but in this case the protective film consists of two layers: ferric oxide is the inner layer next to the metal; the outer layer is ferric phosphate. Some ascribe passivity to the presence of a monolayer of oxygen molecules adsorbed on the surface of the metal.

It was shown in Chapter I that the effect of impressing a positive electrical potential on a metal is to withdraw electrons, thus causing the metal to go into solution. That is, a positive potential tends to enhance oxidation half-reactions.

$$Fe = Fe^{++} + 2e^- \qquad (V\text{-}1)$$

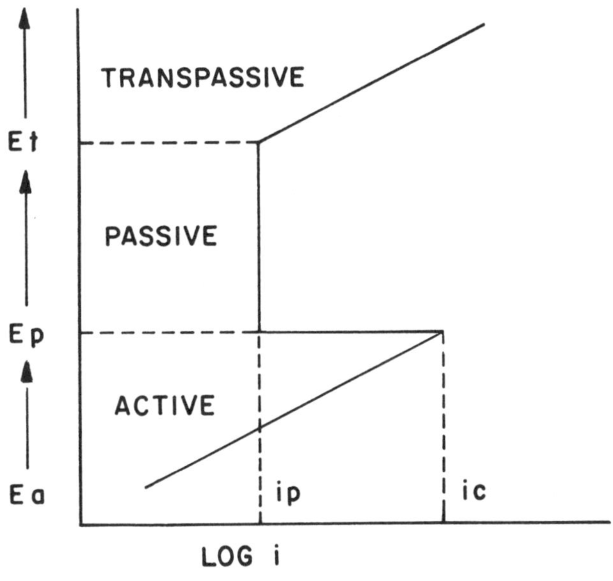

Fig. V.1

E_a = active potential range
E_p = passive potential range
E_t = transpassive potential range
i_c = critical current
i_p = passive current

When an anodic polarization curve is plotted of a metal such as iron, which undergoes an active-passive transition, the shape of the curve is more-or-less as shown in Fig. V.1. In investigating the mechanism of passivity of nickel in acidic solution, Bockris, et al.[1] found that as an increasingly positive potential is impressed, the corrosion current increases. As the potential approaches E_p and the current approaches i_c, a film of oxide begins to form on the metal surface, thickening to a maximum of about 60 Å at i_c. Then, as the current begins to decrease, there is a corresponding increase in the electronic conductivity of the film and a decrease in its ionic conductivity, which obstructs the flow of anions needed to complete the internal circuit. Continuing to increase the potential does not change the current, which remains essentially constant at i_p, but when the potential reaches E_t the current begins to increase as the metal begins to corrode. Through the range of potentials from E_p to E_t (for iron from about 0.6 to 1.6 V) the metal is protected from corrosion. The work of Bockris, et al.[1] suggests that the essential cause of passivity is neither a monolayer of oxygen nor a multilayer of oxide, but rather it is the transformation in the multilayer from ionic to electronic conductance at the passivation potential. Foley, et al.,[2] in electron diffraction studies of oxide films on iron, found that the film present in the passive region is γ-Fe_2O_3, a poor ionic conductor, while in the active and transpassive ranges the film is Fe_3O_4, a relatively good ionic conductor. Although the compositions of the films are generally independent of pH, West[3] points out that, in the presence of acids, reductive dissolution of metal by a passivating film occurs.

$$Fe + Fe_2O_3 + 6H^+ = 3Fe^{++} + 3H_2O \qquad E^\circ = 1.17\ V \qquad (V\text{-}2)$$

The net result is that ferric ions in the lattice are either brought into solution as ferrous ions or reduced to the ferrous state in the lattice, either process producing vacancies in the surface that facilitate the dissolution of iron. Iron made passive in one normal sulfuric acid by an impressed potential, for example, begins to corrode almost immediately (1-3 s) when the potential is removed. Also, passivity may be destroyed if steel is connected to magnesium sacrificial anodes in water having high specific conductance, e.g., sea water.

As noted above, the minimum passivating potential of iron is +0.6 V, so oxidizing agents with potentials above this value passivate the surface of iron. The more important of these are chromate and nitrite. Examination under a scanning electron microscope reveals that a specimen of carbon steel pretreated for 30 min at 130–140°F with a solution containing 1000 ppm CrO_4 –is uniformly covered by an oxide film 90–100 Å thick, composed of 90 percent Fe_2O_3 and 10 percent Cr_2O_3, the concentration of chromium at the surface being 7 $\mu g/cm^2$.* Lumsden and Szklarska-Smialowska[4] found that a passive film formed on iron, 45–58 Å thick, upon immersing specimens in $0.01N$ Na_2CrO_4 for 1 h at pH 7.

Nitrite passivates iron only above pH 6, below which value it is unstable with respect to decomposition to NO and NO_2.[5]

$$2HNO_2 = NO_2 + NO + H_2O \qquad \text{(V-3)}$$

Passivating reactions also include:

$$2Fe + H_2O + 3/2O_2 = Fe_2O_3 \cdot H_2O \qquad \text{(V-4)}$$

$$2CrO_4^{--} + 2H_2O + 2Fe = Cr_2O_3 + Fe_2O_3 + 4OH^- \qquad \text{(V-5)}$$

$$2HNO_2 + 2Fe = Fe_2O_3 + H_2O + \tfrac{1}{2}N_2 \qquad \text{(V-6)}$$

Passive films discussed thus far are thin, 40–100 Å thick, and thus, invisible. When phosphates are used to facilitate the adsorption of oxygen, which is ultimately reduced as shown in Eq. (V-4), much thicker films are formed whose physical properties are dependent on pH. For instance, Lumsden and Szklarska-Smialowska[4] found that it took more than 16 h to form a film 12 Å thick in $0.1N$ Na_3PO_4 at pH 12.3. The film contained only a small proportion of phosphate, which was mainly on the outer surface. In $0.1N$ Na_2HPO_4, however, at pH 9.1, a film approximately 200 Å thick was formed, composed of two layers, the inner layer next to the metal was γ-Fe_2O_3 and the outer layer was $FePO_4$. In the presence of zinc, Trillat and Saison[6] found that next to the metal the film was of the same composition as obtained with other methods of passivation, namely, γ-Fe_2O_3, but covering this were lamellae of $Zn_3(PO_4)_2 \cdot 4H_2O$ upon which was superposed an epitaxial monoclinic growth of $(Zn,Fe)_3(PO_4)_2 \cdot 4H_2O$; they also saw evidence of primary and tertiary ferrous phosphates, i.e., $Fe(H_2PO_4)_2$ and $Fe_3(PO_4)_2 \cdot H_2O$. Phosphate protective coatings are, in effect, scales that form over a layer of γ-Fe_2O_3 providing a physical barrier. They are usually thick enough to be visible regardless of whether formed by polyphosphates, pyrophosphates, or combinations of orthophosphates; for complete passivation of iron in aerated water the total concentration of phosphate, as PO_4^{-3}, must be greater than 1500 ppm; usually stronger solutions are specified to hasten the reaction, which is slow at best. Crystalline phosphate films are highly permeable to ions and, by themselves,

* J. Q. Adams, Chevron Research Company, Richmond, CA 94802. Private communication.

provide less protection than the thinner, amorphous γ-Fe_2O_3.

Cartledge and Sympson[7] have established the existence of a Flade[8] potential on iron passivated with chromate ion. This potential, E_F, occurs at an equilibrium in the transition of iron from the passive to the active state. At this equilibrium potential, which depends upon the oxide film and the pH, no current passes through the film. The Flade potential of iron is given by the following expression.

$$E_F = -0.63 + 0.059\,\text{pH} \qquad \text{(V-7)}$$

The negative potential, –0.63 V, indicates a pronounced tendency for passivity to decay, and this is hastened by chloride ions, reducing agents, high temperature, surface scratches, and strong magnetic fields. A fundamental step in the corrosion of iron is the hydration of the ferrous ions produced by the oxidation of iron. The intense electrical field surrounding these ions attracts water molecules; the large energy of hydration strongly favors corrosion. The formation of the passivating film displaces water molecules from the metal surface and shields it from further hydration. In the presence of an appreciable concentration of chloride ions, however, it is difficult to passivate iron effectively because of the tendency of chloride ions to penetrate small defects in the passive film and to form complexes, such as $FeCl_6^{-3}$, with iron. It is for this reason also that passivity decays quickly in chloride solutions.

V.2 PASSIVATING PROCEDURES

After acid-cleaning any steel vessel it is essential that residual acid be neutralized and that the surfaces be passivated by applying a suitable solution before returning the equipment to service. It is also necessary to deoil and lightly acid-clean new steel equipment to remove mill scale and prepare the surface for passivation; this is often neglected in retubing heat exchangers. Following cleaning with an acid, rinsing with water, and neutralizing—usually with sodium carbonate—the highly reactive surface tends to flash-rust. The formation of this type of ferric oxide can be prevented by nitrogen-blanketing or by adding sodium gluconate or triethanolamine to the passivating solution. Both of these compounds form very stable complexes with ferric ion and thus prevent the formation of rust, α-Fe_2O_3, without interfering with the development of the passivating film of γ-Fe_2O_3.

Numerous passivating solutions have been proposed[9,10,11] including sodium hydroxide and sodium carbonate, various phosphates and polyphosphates, sodium nitrite, and chromate, among others. Table V-1 contains a number of procedures that have proven satisfactory for passivating boilers subsequent to chemical cleaning. Methods 3-8 in the table are carried out at 150 psi using natural circulation.

TABLE V-1

Passivating Solutions

	Solutions							
Components	1	2	3	4	5	6	7	8
NaOH	–	–	1.0	1.0	0.2	–	–	–
Na_2CO_3	–	1.5	–	–	0.5	1.0	–	0.25
Na_3PO_4	–	–	–	0.125	–	0.125	0.125	–
Na_2HPO_4	0.25	–	–	–	–	–	–	0.25
NaH_2PO_4	0.25	–	–	–	–	–	–	0.25
$NaNO_2$	0.25	0.25	–	–	–	–	–	0.5
Sodium gluconate	–	0.25	–	–	–	–	–	–
Triethanolamine	–	–	–	0.2	0.1	–	–	–
N_2H_4	–	–	0.05	–	–	–	0.02	–
Nitrogen blanket	Yes	No	No	No	No	Yes	Yes	Yes
Temperature, °F	200	200	350	350	350	350	350	200
Duration, hours	1*	4**	6	3-6	3-4	3-6	3-6	3-6
Stainless steels	Yes	Yes	No	No	Yes	Yes	Yes	Yes
Reference	(9)	–	(10)	(10)	(10)	(10)	(10)	(10)

* Circulate 15 min; soak 30 min; drain.

** Add sodium carbonate and sodium gluconate, circulate 2 h at 200°F. Add sodium nitrite, circulate 2 h at 200°F. Drain.

Many boiler operators are reluctant to use sodium hydroxide in high-pressure boilers (2000 psi +) because of the possibility of inducing stress corrosion cracking in stainless steel tubes. Three passivating formulations are discussed here, each suited for a particular purpose, that perform well and present minimum difficulties in waste disposal. Passivating is accomplished by circulating the solution through equipment such as boilers, reactors, and heat exchangers; bundles of tubes are sometimes passivated by dipping in a vat, if their configuration is such that they can be readily drained.

V.2.a Phosphate and Nitrite

Duncan, et al.[9] have found that a solution containing 0.25 percent each of Na_2HPO_4, NaH_2PO_4, and $NaNO_2$ provides very even passivation in acid-cleaned steam generators; it is the formulation of choice for boilers, columns, pressure vessels, and heat transfer equipment. Without leaving a visible film on the surface, this combination passivates iron, aluminum, nickel, and stainless steels by lowering the critical current. In general, it is unnecessary to passivate stainless steels containing more than about 12 percent of chromium for, because of the affinity of their surfaces for oxygen, they are self-passivating. Aluminum, too, is protected by a naturally occurring surface film of Al_2O_3, which in air, is only a few angstroms thick, although in aqueous solution the film thickens considerably.

Banks and Martin[10] observed that in using sodium nitrite at elevated temperature (350°F) the passive film was weakened by the heat. At 150–200°F, however, a very even and stable passivation is achieved. This is especially important in passivating high-carbon steels, as cementite, Fe_3C, passivates more readily than the iron phase, thus acting as a cathode. When the passive film on steel is uneven or unstable rapid pitting can ensue.

When rinsing and draining after acid-cleaning a boiler, a nitrogen blanket is necessary to prevent flash rusting of the active steel surfaces upon exposure to air. Alternatively, 0.25 percent of sodium gluconate may be added to the passivating formulation to dissolve any rust formed during rinsing and neutralizing. If this is done, however, it is essential to check the pH of the solution to be sure it remains in the range of 6–7; the pH of the phosphate–nitrite combination without gluconate is buffered at about 6.3. The solution should be heated to 200°F, then circulated through the unit for about 15 min, followed by 30 min of soaking before it is drained. The procedure is described in greater detail in Chapter VII.

V.2.b Chromate

A solution of chromate provides exceptionally rapid passivation of iron, aluminum, cadmium, lead, magnesium, copper, and zinc; the film formed on steel is stable for several days, even if the metal is wet. About 30 min are required for a protective film to develop on a steel surface exposed to a cold, aerated solution containing 1000 ppm of chromate. If the solution contains no oxygen, film formation takes some four times longer. In a similar solution, heated to 140°F, a film 90-100 Å thick is complete in 5 min, and does not thicken further with longer treatment. Chromate is the obvious choice for passivating heat exchangers cooled by chromate-treated cooling water, as the solution can be flushed into the recirculating cooling water when a passivated exchanger is returned to service, thus benefitting the entire system while at the same time avoiding a serious waste disposal problem. It is usually convenient to use a regular water-treating inhibitor formulation for passivating provided its components are all soluble at the concentration and pH used for passivating, in the case of chromate, 6-7. Suppose, for example, that the water-treating inhibitor contains 51 percent CrO_4^{--} and 6.9 percent Zn^{++}. If a solution is prepared containing 1000 ppm of CrO_4^{--}, then the concentration of Zn^{++} will be 135 ppm, because the ratio of chromate to zinc ion in both solutions is 7.4. The solubility product of $Zn(OH)_2$ is $K_{sp} = 1.2 \times 10^{-17}$, from which the maximum concentration of Zn^{++} that can remain in solution can be calculated at any value of pH. Suppose the pH is 6.0. Then,

$$\begin{aligned}
(H^+) &= 10^{-6} \\
(OH^-) &= 10^{-14}/10^{-6} \\
&= 10^{-8} \\
(Zn^{++}) &= 1.2 \times 10^{-17}/(10^{-8})^2 \\
&= 0.12 \text{ mol/l} \\
0.12 \text{ mol/l} \times 65 \text{ g/mol} &= 7.8 \text{ g/l} \\
&\cong 0.78\%
\end{aligned}$$

As the concentration of zinc ion in the passivating solution will be only ≈ 0.014 percent, all of it will remain in solution at pH 6. Note that at pH 7, on the other hand, the concentration that can remain in solution is one-hundredth that at pH 6, consequently more than half of the amount added with the chromate will be precipitated.

To determine the effect of high-pressure water jetting on passive films, a 9000-psi jet of water was passed over a chromate-treated specimen at the rate of 1 fps. The original thickness of the film was 94 Å with a surface concentration of chromium of 7 $\mu g/cm^2$. After being subjected to the jet the thickness was 60 Å with a surface concentration of chromium of 4 $\mu g/cm^2$, i.e., some 35 percent of the film was stripped. Although it might be inferred from these results that passivation is unnecessary after water-jetting, it must be done, as there is no assurance that scratches or unfilmed areas are not present.

In practice, a solution containing 1000 ppm of CrO_4^{--}, at a pH of 6–7 is heated to 130–140°F, and pumped into the heat exchanger to be passivated. After 30 min of circulating or soaking, the exchanger is reconnected to the headers, and the chromate solution is flushed into the recirculating cooling water.

V.2.c Polyphosphate

The presence of polyphosphate facilitates the passivation of iron, aluminum, cadmium, copper, lead, and zinc by oxygen. Oxygen being the passivating agent, it is essential that the solution be sufficiently aerated to achieve complete passivity. This is ensured by prolonging the treatment, a decided disadvantage of this method, especially when it is not possible to attain a high temperature. Any polyphosphate can be used, but sodium hexametaphosphate is least expensive, and is available powdered, crushed, or in the form of glassy chips; the powdered product has the highest solution rate. A solution is prepared containing 1 percent of polyphosphate (as PO_4^{-3}), the pH is adjusted to 5.5–6.5, and the solution is heated and circulated. The length of time required to passivate steel depends upon the temperature of the solution: 1 h at 200°F; 4 h at 180°F; 8 h at 140°F; and 24–48 h at 60–70°F.

The protective coating consists of two components: a film of $\gamma\text{-}Fe_2O_3$ on the metal surface overlain with a thick, visible diffusion barrier of $FePO_4$, resulting from the hydrolysis of polyphosphate to orthophosphate. Note that this hydrolysis decreases the pH of the passivating solution.

$$PO_3^- + H_2O = H_2PO_4^- \qquad \text{(V-8)}$$

As solutions of polyphosphate are rather soapy and prone to foam, particularly if a wetting agent is used, a defoamer may be advisable. These are usually included in proprietary formulations. Polyphosphate passivation is most appropriate for heat exchangers in which nonchromate-treated cooling water is circulated, and in circumstances under which it is possible to heat the solution to 200°F so that the process can be completed in a reasonable length of time.

References

(1) J. O'M. Bockris, Reddy, A. K. N., and Rao, B. 1966. An ellipsometric determination of the mechanism of passivity of nickel. *J. Electrochem. Soc.* 113 (11): 1133.

(2) Foley, C. L., Kruger, J., and Bechtoldt, C. J. 1967. Electron diffraction studies of active, passive, and transpassive oxide films formed on iron. *J. Electrochem. Soc.* 114 (10): 994.

(3) West, J. M. 1971. *Electrodeposition and corrosion processes.* London: Van Nostrand Reinhold Co., Ltd.

(4) Lumsden, J. B. and Szklarska-Simalowska, Z. 1978. The properties of films formed on iron exposed to inhibitive solutions. *Corrosion* 34 (5): 169.

(5) Yost, D. M. and Russell, H., Jr. 1944. *Systematic inorganic chemistry.* New York: Prentice-Hall, Inc.

(6) Trillat, J. J. and Saison, J. 1961. Etude des couches de phosphation au moyen de rayons X et de la diffraction électronique. *Mem. Sci. Rev. Met.* 58 (7): 496.

(7) Cartledge, G. H. and Sympson, R. F. 1957. The existence of a Flade potential on iron inhibited by ions of the XO_4^{n-} type. *J. Phys. Chem.* 61: 973.

(8) Flade, F. 1911. Beiträge zur Kenntnis der Passivität. *Z. Physik. Chem.* 76: 513.

(9) Duncan, F. R., Fyock, D. H., Baldwin, L. V., and Hopkins, R. D. 1972. Preoperational cleaning of high pressure boilers of General Public Utilities Corporation. *Proc. Int'l. Water Conf., Eng. Soc. W. Pa.* 33: 97.

(10) Banks, W. P. and Martin, L. D. Dec., 1977. High temperature passivation in natural circulation boilers. *Materials Performance* 16: 21.

(11) Hantelman, W. D. 1972. Ten years experience in preoperational chemical cleaning of power plant steam generating systems. *Proc. Int'l. Water Conf., Eng. Soc. W. Pa.* 33: 101.

VI

Planning and Contracting

Chemical cleaning is an effective method for restoring and prolonging the operating efficiency of industrial equipment, especially with respect to heat transfer. The escalating cost of energy provides a strong incentive for the application of this technique for improving the performance of boilers, heat exchangers, surface condensers, reaction coolers, and other devices used to transfer heat. In addition to reducing expenditure for energy, chemical cleaning may also provide the benefits of longer unit runs and the saving of time on shutdowns.

Among the items that must be carefully considered in planning and executing a chemical cleaning are preliminary testing of fouling material, inspection of the condition of equipment to be cleaned, possible hazards to be encountered, and the troublesome issue of waste disposal. These and other subjects are discussed in a general way in this chapter, then in the succeeding four chapters detailed procedures are suggested for cleaning specific types of industrial equipment.

VI.1 PRELIMINARY TESTING AND INSPECTION

In order to select a solvent for dissolving fouling material the chemical cleaning contractor needs to know at least its approximate composition (which can be determined by a chemical or physical analysis), and its solubility in selected solvents. The primary concerns are the solubility of the material to be removed and the metallurgical composition of the equipment to be cleaned. For obvious reasons, it is essential that the solubility and analytical work be performed on samples that are representative of the bulk of the fouling material, although in many instances this is a difficult prerequisite to fulfill. The preliminary examination and description of the sample should include its appearance, consistency, flammability, odor, and behavior with a magnet.

Scales are usually sampled by scraping with a sharp steel blade, taking care not to introduce any metallic chips from the blade or substrate. Laminar scales, such as occur in furnaces, may be segregated into layers at the site of sampling, if desired, but if it is possible to dislodge a fairly large piece, it is more convenient to separate the laminae in the laboratory.

Wet sludges can be separated from the bulk of their water by straining through cloth, dipping with a sieve, or collecting on paper towels. Wet samples and those from corrosive environments should not be stored in cans for the container is rapidly attacked and contaminates the sample. For the majority of samples a wide-mouth Pyrex bottle is a suitable container.

The sampling and storage of material from hydrofluoric acid systems presents special problems. In large plants it may be worthwhile to have a few monel cans made for sampling, although they are expensive. Samples of this type should be clearly marked, preferably with tags or labels of distinctive color, and handled with appropriate caution when received in the laboratory. Corrosive systems are generally washed with water and steamed before being opened for inspection and maintenance and this, of course, may alter the deposit. The formation of acid salts (e.g., KHF_2) is common in some sections of alkylation plants using hydrofluoric acid as catalyst, but these are ordinarily dissolved in the preliminary washing operation.

The original sample is customarily collected under the supervision of a process engineer, while the results of the analysis are interpreted by the analyst or the cleaning contractor. In general, determinations of great accuracy are unnecessary because the nature of the usual deposit is such that much time and effort would have to be expended to obtain a truly representative sample and, anyway, solubility is the principal concern here. Rapid methods that give reasonably accurate results, say within 5 percent of the amount actually present, are therefore in order.[(1)] Using this approach, the results obtained are adequate for the purpose of the contractor, while at the same time the cost of the analysis is not excessive.

Spectrographic analysis is useful, if equipment is available, for determining the presence or absence of metallic elements.[(2)] Routinely, it is usually not possible to make quantitative determinations of the principal metallic components more precisely than ± 10–15 percent. In general, the components that are present in concentrations below 1 percent are more accurately determined spectrographically, but these are ordinarily of slight interest in the analysis of deposits preliminary to chemical cleaning. On the other hand, distortion from metals present in high concentration may occur because of overlapping spectral lines; these produce indeterminate errors in the determination of elements present in small amounts. As a rule, the relative values in a spectrographic analysis are better than the absolute values.

X-ray diffraction has been applied with some success to crystalline, water-formed deposits for the identification of chemical compounds.[(3)] When many different crystal forms are present, however, the interpretation of the diffraction patterns is extremely complicated, and for this reason the method is not especially applicable to corrosion deposits or those containing a variety of elements.

Deposits are formed either as a result of attack on a vessel or through deposition of material being processed in it; those elements that occur in various steels and alloys in small amounts and some used only for special purposes can usually be eliminated from consideration. Table VI-1 includes the elements most often encountered in process equipment, their frequency of occurrence, and the type of deposit in which they are found. The frequencies indicated are for the occurrence of the particular element in concentrations exceeding 1 percent. It is emphasized that the frequencies are general and exception could be taken to some of the classifications. Thus, copper in water-formed deposits is occasional in a large number of samples; it is common, however, in utility boilers and in marine and industrial boilers using evaporated make-up water. Special cases exist, of course, where any of the components would be commonly found.

In most instances a complete chemical analysis is unnecessary, but solubility tests should always be made giving due consideration to the materials of construction (as discussed in Chapter II and summarized in Table II-8), the temperature that is attainable, and whether or not the cleaning solution can be circulated. During the solubility tests it is possible to ascertain whether the evolution of a toxic gas will be likely during the cleaning process. Also, the presence of large amounts of carbonate can lead to overpressuring by carbon dioxide liberated upon acidification. For these reasons, testing must be done cautiously in a fume hood and the details should be carefully recorded so that the cleaning can be carried out safely and effectively.

TABLE VI-1

Frequency of Occurrence of Components of Deposits

Component	*Corrosive attack*	*Furnaces*	*Water-formed*
SiO_2*	I	C	C
Fe	C	C	C
V*	R	C	R
Mo*	R	R	R
Pb	C	I	R
Cu	C	R	I
Al*	I	C	C
Cr*	C	I	C
Ni*	C	C	R
Zn	I	R	I
Ca	I	C	C
Mg*	I	C	C
Na	I	C	C
C	C	C	C
CO_3^{--}	R	I	C
SO_4^{--}	C	C	C
PO_4^{-3}	I	R	C
Cl^-	I	I	I
F^-	I	R	R
S^{--}	C	I	R

C = common; I = infrequent; R = rare.

* These elements are frequently used as catalysts in petroleum refining.

A further hazard that must be considered is the possible presence of pyrophoric iron sulfide, which is particularly likely to be found in natural gas lines, petroleum distillation units, preheat furnaces, and crude oil tankage. Hughes, et al.[4] point out that α-FeOOH, the common form of rust, forms an especially reactive pyrophore with hydrogen sulfide, which reoxidizes to Fe_2O_3 when exposed to air.

$$Fe_2O_3 + 3H_2S = 2FeS + S + H_2O \qquad \text{(VI-1)}$$

$$4FeS + 3O_2 = 2Fe_2O_3 + 4S \qquad \text{(VI-2)}$$

As the reactions represented by Eq. (VI-1) and Eq. (VI-2) are both exothermic, they are thermally self-sustaining so when the pyrophoric sulfide is exposed to air the surface immediately begins to exhibit localized glowing and sparking. Highly reactive pyrophoric iron sulfide can also form in aqueous solutions containing about 400 ppm of H_2S. The material undergoes controlled oxidation as it dries, but if dried before being exposed to air it is subsequently intensely reactive. The pyrophore has a metallic luster that gives warning of the presence of a possible source of ignition.

The general objectives in chemical cleaning are to maximize the return on the cost of cleaning by improving the efficiency of a process, to remove organic coatings and mill scale from new equipment before initial start up after a major retubing, and to remove scales and deposits before irreversible damage is done to the metal beneath them. The need for chemical cleaning is determined by internal inspection of equipment, such as boilers and surface condensers, and by external inspection, where appropriate as, for example, to observe bulging or bagging of boiler tubes, or exfoliation of cupronickel superheater and feed water preheater tubes. Thorough inspections, using the best method available, should be made both before and after chemical cleaning. This permits comparison between the condition of the equipment at these times and when it was new; this may indicate either process upsets or improper cleaning methods. For instance, eddy current testing can detect discontinuities and defects in the walls of a vessel, or the existence of a layer of scale on an internal surface. These affect the flow of eddy currents which in turn produces variations in the impedance of an eddy current-inducing coil.[5] When compared with a detailed eddy current map of the original equipment, made before cleaning, significant changes can be revealed.

Ultrasonic inspection[6] can detect thinning of tubes within ± 10 mils if the base thickness has been previously measured, or is otherwise known. The ultrasonic pulse is attenuated in the presence of pits or fissures caused by hydrogen attack or tertiary creep; the method has the advantage that large areas can be quickly scanned. Radiography[7] (x-ray photography) can be used to confirm internal pitting, loss of metal, the presence of scale, and the condition of welds. Optical methods of inspection include fiber optics devices and miniature television cameras that permit visual inspection of the insides of boiler tubes and similar otherwise inaccessible areas. Internal inspection of a boiler for pitting and cracking often must be preceded by power brushing or water jetting to prepare the surface for observation.

When cleaning steel equipment with properly inhibited hydrochloric acid, the corrosion rate should not exceed 0.8 mg/cm^2/h (0.04 mil/h) of uniform etching. If the acid is left in the vessel too long, at too high a temperature or circulation rate, however, extensive pitting will occur that is particularly intense in stressed areas and in pre-existing pits. Heavy corrosion deposits in the mud drum of a boiler, and circumferential corrosion in the highly stressed areas where tube ends are rolled into the drum are also characteristic of acidic attack. In the presence of ferric ion, extensive pitting is common, and cracking may also be seen in weld seams. Pits caused by acidic attack are, in general, empty; a pit containing ferric oxide after acid cleaning was very likely formed before cleaning. Circumferential cracking of tube walls is often attributable to low pH of boiler feed water, as are intensive pitting, heavy scaling, and frequent failures of downcomer and waterwall tubes.

Copper is often seen in boilers, either plated on the surface of tubes or mixed with magnetite; occasionally a layer of metallic copper is seen overlying a protective film of magnetite, indicating that the copper was introduced in the feed water. Scratches are sometimes mechanically produced during the manufacture of boiler tubes. Although these often resemble cracks, they can be characterized by making a photomicrograph and examining it for the presence of deformed grains adjacent to the scratches. Lack of deformation suggests that the scratches were made while the metal was hot, i.e., during manufacture. A similar phenomenon has been observed caused by using aerated rinse water or leaving a boiler wet for several days after cleaning. What appear to the unaided eye as cracks or grooves are seen under magnification to be strings of pits, presumably along draw marks, caused by oxygen. Chemical cleaning appears to enhance this type of attack, but exposure to oxygen is the proximate cause.

VI.2 CONTRACTING

The execution of a chemical cleaning contract is a legal procedure in which a contractor and a client specify terms of agreement on how, where, and when one or more pieces of equipment are to be cleaned. Firm, lump-sum bids may be solicited in which the contractor undertakes to carry out an assignment for a fixed sum of money. In this type of agreement it is essential that every detail of the procedure be specified, including mechanical work to be done by the client, for any extra work or delays not attributable to the contractor require negotiated changes in the total cost. Extra charges are incurred if additional equipment must be provided to handle contingencies not anticipated by the client, lines must be relocated, time is lost while the client installs flanges, blinds, or valves, and if excess waste must be disposed of because of changes in procedure requested by the client subsequent to the signing of the contract. Contractors invariably over-bid lump-sum contracts to protect themselves from unavoidable over-runs.

Stander[8] points out that a great deal of the client's time is required to evaluate bids for individual jobs and suggests that term contracts are more satisfactory for routine work. In this arrangement the contractor initially estimates the cost of labor, materials, and equipment, then submits billings by unit prices, such as dollars per manhour, dollars per 100 gal of hazardous waste handled, and dollars per hour for equipment rental.

Before assigning large industrial jobs it is customary for a client to convene a bid meeting with interested contractors to discuss the details of the undertaking. At this meeting the name, location, service, materials of construction, volume, and pressure rating of each unit to be cleaned are disclosed, along with the desired starting and completion dates. An inspection at the job site is also made at this time. It is helpful to the contractor if the nature of the fouling and an estimate of the quantity to be removed can be given. This is usually impossible as units are normally not opened before chemical cleaning, but a history of operation, especially the analyses of samples removed during previous shutdowns, is valuable in selecting an appropriate solvent.

The cleaning procedure and solvent may be either the client's preference or that of the contractor. In some instances the client specifies the solvent to be used, the inhibitor, the method of heating the solvent and its temperature, the duration of exposure of the unit to the inhibited solvent, the sampling and testing schedule, as well as the exact procedures for rinsing, neutralizing, and passivating. Regardless of who proposes the procedure, however, both parties must agree on how it is to be carried out, and this must appear in the contract. If the contractor considers a procedure proposed by the client or some third party, such as a consultant, inappropriate he may insist that a liability release be signed. In this event it is usually advisable to find another consultant. Other items to be covered in the agreement include the supply and handling of utilities, who is to dispose of the waste, whether on or off site, records to be kept by the contractor, preferably on forms supplied by the client, and clean up after the job is completed. Some form of insurance may be required, and the client must assure himself of the competency of the contractor, particularly with respect to safe practices including compliance with Federal, state, county, and city codes, laws, rules, regulations, and industrial standards, as well as the *Occupational Safety and Health Act*.

Chemical cleaning can be carried out safely and effectively only if there is close cooperation between the client and the contractor. For this reason it is desirable to appoint a coordinator to deal with the contractor and the operator of the equipment to be cleaned. His function is to identify hazards, specify safety equipment, issue instructions for piping changes, anticipate staging requirements, see that interference with regular operations is minimized, and follow the course of the work to ensure that it is carried out as stipulated in the contract. When the job is completed the coordinator checks the contractor's charges, verifies that any extra charges are legitimate, and certifies that the work was performed according to the terms of the contract.

VI.3 PLANNING FOR CHEMICAL CLEANING

The first point to be considered in planning a chemical cleaning job is the reason for doing it at all. Acid cleanings are carried out to remove mill scale from new equipment or to remove fouling from operating equipment. The former application is a routine preoperational procedure for steam generators, heat exchangers, and other equipment that is designed for aqueous service. Operational cleaning is undertaken to remove fouling that is in some way interfering with the efficiency of an operating unit, manifested by decreasing heat transfer coefficients, increasing pressure differentials across equipment, restricted flow, or increasing pumping resistance. The objective is to remove the fouling without damaging the equipment, at minimum cost. As applied to chemical cleaning "cost-effectiveness" is an elusive concept visualized somehow in terms of benefits obtained for the expenditure of a certain amount of money.

Major factors in the cost of chemical cleaning are down time, production loss while cleaning, the rental of special apparatus including travel and standby charges, the cost of labor and materials, and the now almost prohibitive charges for disposing of hazardous wastes. Offsetting these direct costs are those of operating at lowered efficiency, the cost of energy wasted in impaired heat transfer processes, and the real possibility of irreversibly damaging process units by running to failure. By careful planning and consulting with the chemical cleaning contractor all these factors can be integrated to maximize the return on cleaning costs while extending the life of equipment. Lawson[9] has devised an intriguing quantitative approach to optimizing the economic aspect of chemical cleaning.

In formulating a plan, provision msut be made for isolating the work area, spotting trucks in convenient locations, obtaining necessary permits, avoiding interference with other work in the area, ensuring the availability of utilities, and for the emergency disposal of cleaning solutions. Safety meetings must be scheduled with operations personnel and the contractor's crew to anticipate the greatest hazards such as the release of flammable liquids or toxic gases, and to provide safeguards against them. The ultimate disposal of waste solutions should be on site, if possible, because of the great expense of off site disposal incurred to some extent by the necessity of transporting a large volume of water. Every effort should be made by the client to be ready to proceed when the contractor and his equipment arrive, thus minimizing standby and overtime charges. Also, in comtemplating a chemical cleaning, alternatives should be considered, no matter how unpromising they may appear at first glance. There are situations, for instance, in which it is less expensive to retube a heat exchanger than to chemically clean it, and it is often possible to justify spare tube bundles that can be quickly exchanged for fouled ones; the latter can then be cleaned later by a less expensive method, such as water jetting.

The volume of hazardous wastes generated can be minimized by proper selection of cleaning methods. High pressure water jetting is usually the method of choice for cleaning either the tube side or the shell side of tube bundles in heat exchangers. The tubes are cleared by pushing flexible lances attached to a source of high pressure water through them. The outside of tube bundles can be cleaned by removing the bundle from its shell and cleaning it with a bundle blaster, a machine that automatically rotates the bundle while jets of water at up to 10,000 psi move longitudinally along the tubes. Virtually no hazardous waste is generated by water jetting. Triangularly pitched tube bundles, however, cannot be cleaned by this procedure as the interior tubes are not accessible to the water jets. Acidic foam is appropriate for cleaning large equipment such as columns and surface condensers as only about 10 percent of the volume of acidic waste is produced as compared to cascading and, in addition, more surface is reached by foam. Nitrogen/abrasive cleaning is suitable for pipelines and furnace tubes, provided enough footage is to be cleaned; set-up cost is about the same for 100 ft as for 1000 ft of pipe. This method of cleaning gives the best results with piping of uniform diameter in which the pipe and fouling are dry and oil-free. Walnut hulls are as effective an abrasive as is sand and they are less damaging to valve trim; steel shot is superior to both. Line moling is useful for piping containing oily deposits, but it is necessary to install flanges every 200 ft. Moling is carried out at a pressure of 4000-7000 psi, with a water-flow rate of 15-20 gpm. On-stream cleaning schemes using chelants have been suggested for low-to-moderate-pressure boilers and for cooling systems; they produce little or no hazardous waste, but they are expensive, particularly when applied to large recirculating cooling water systems.

Before turning a piece of equipment over to a contractor for cleaning, the client repairs any leaks and isolates the unit from steam, water, gas, petroleum, or chemical lines. Generally, the client installs 2-in. piping connections for filling, circulating, and venting. A block valve is also provided on the lower connection and, if toxic gases may be released, rigid piping should be installed to connect the unit to a caustic or other scrubber or, alternatively, to a flare or gas recovery system. Units requiring frequent cleaning are often fitted with permanent but removable piping, with flanges designed for easy assembling and dismantling. Fittings are constructed of welded steel piping. Block valves on heat exchangers in cooling water service should be slip-blinded if permanent cleaning connections are provided; if not, spools must be removed to provide access to the unit. Blinding should not be omitted for expedience, as cleaning solution is likely to damage valve trim, and leak into the cooling water. Having taken account of the foregoing mechanical considerations, the cleaning process itself is now ready to be addressed.

The more important factors affecting the outcome of a chemical cleaning procedure include the nature of the fouling; the metallurgy of the equipment to be cleaned; the temperature, concentration, and velocity of the selected solvent; the corrosion inhibitor used; and the rinsing, neutralizing, and passivating steps carried out in preparation for returning the unit to service. Many different types of fouling deposits are described in Chapter I and methods for their analysis have been outlined in Section VI.1. In general, if some solvent, usually an acid, dissolves 40-50 percent of a deposit sample that solvent will provide an adequate cleaning of a piece of industrial equipment, and more often than not, will perform better in the field than in the laboratory. In subsequent chapters specific types of deposits and solvents are discussed. Here it is sufficient to repeat that deposits coated or mixed with hydrocarbons must be deoiled before attempting to dissolve the inorganic components in acidic solvents. Heavy deoiling is best done with an alkaline emulsion, as described on p. 40; light deoiling can be accomplished by applying a dilute solution of trisodium phosphate, as discussed on pp. 35-36.

Often the solvent that proves most effective for dissolving fouling is incompatible with the metal of which the fouled unit is constructed. Many of these metallurgical considerations are summarized in Table II-8, p. 62. It is not at all uncommon to find brass tubes rolled into a steel tubesheet, and although stainless steels are eminently unsatisfactory for cooling water service, they are commonly encountered in coolers and heat exchangers in chemical manufacturing and hydrogen sulfide recovery plants. Special care must be exerted in planning a chemical cleaning job to take into account the materials of construction, including special gaskets and valve trim (seats, disks, and stems), which are often fabricated from unusual alloys. Such valves should be slip-blinded, or replaced with spools, as already mentioned, and inert gaskets should be substituted during cleaning.

Ideally, the solvent used for cleaning should be highly efficient in removing fouling, minimally corrosive to equipment, and easily disposed of after use. The planner must somehow arrive at a reasonable compromise among these conflicting desiderata. As one of the more important factors in the efficacy of a cleaning solvent is its temperature, it is important to ensure that the desired temperature can be attained with existing utilities. If not, some sort of heater must be supplied, either by the client or by the contractor. In most instances sufficient heat is obtained by injecting 150-psi steam into the solvent line downstream of the circulating pump. If only lower pressure steam is available, a heat exchanger should be used to avoid excessive dilution of the solvent by condensate. Care must be taken not to exceed the maximum temperature specified for the corrosion inhibitor; the corrosion rate of steel in inhibited 5 percent hydrochloric acid, for example, increases very rapidly above 180°F. Corrosion monitoring is advisable while cleaning large or critical units, to avoid serious accidental damage.

It is essential to anticipate the possible release of toxic gases and to provide for their capture. Ferrous sulfide is common in deposits formed in petroleum refining processes; when deposits of this material are acidified copious amounts of hydrogen sulfide are evolved that must be contained. When this gas is expected rigid piping is used for the entire chemical cleaning loop, and caustic scrubbers to absorb the gas are piped to a vent in the upper part of the vessel being cleaned. In this type of operation the inhibited acid is added very slowly to water recirculating in the cleaning loop to avoid too rapid evolution of the extremely hazardous gas.

Other gases that may be encountered in cleaning industrial equipment are hydrogen cyanide, sulfur dioxide, carbon dioxide, hydrogen, and arsine. The latter, which is intensely poisonous, must be absorbed in dilute permanganate solution, as it is unaffected by sodium hydroxide. Arsine is often encountered in alkanolamine gas-sweetening plants in which sodium arsenite has at some time been used as a corrosion inhibitor. Ferrous arsenide formed on steel surfaces is hydrolyzed to arsine in aqueous acidic solutions.

$$Fe_3As_2 + 6H^+ = 3Fe^{++} + 2AsH_3 \qquad \text{(VI-3)}$$

When the cleaning process is complete, as evidenced by no further decrease in acid strength, the acid is drained, the equipment is flushed, often under a nitrogen blanket, after which residual acid is neutralized and, as a final step, the unit is passivated. Some metals do not require passivation and some solvents are themselves passivating agents; when the latter can be used, a significantly smaller volume of liquid waste is produced. For future reference, it is advantageous to document carefully the entire chemical cleaning operation, noting times, temperatures, duration of each step, acid strength, total iron in solution, evolution of gases, and any other pertinent information. This record of experience is often useful in planning and executing cleanings of similar equipment.

Although the desirability of developing and following a specific detailed procedure has been emphasized in the preceding discussion, a reasonable degree of flexibility must be allowed the contractor with respect to heating, nitrogen purging, flow reversal, and time and rate of circulation of cleaning solutions, so that he can most effectively carry out his assignment. Following is a brief checklist that is useful in planning and carrying out a chemical cleaning procedure.

1. Proposed procedure is reviewed for accuracy and completeness.
2. Proposed procedure is approved by operations representative.
3. *Toxic Gas Hazard* is stamped on written procedure or work order, if applicable.
4. Material safety data sheets covering hazardous chemicals that may be encountered are provided.
5. Caustic scrubber is provided, if needed for H_2S.
6. Permanganate scrubber is provided, if needed for AsH_3.

7. Self-contained breathing apparatus is provided.
8. Portable shower is provided.
9. Lighting plant is provided, if needed.
10. Chemicals are supplied by client, if any.
11. Adequate source of water is available.
12. Adequate supply of steam is available.
13. Steam hose or piping is provided, if needed.
14. Contact heater is supplied by client, if required.
15. Rigid piping, if required.
16. Hazardous waste disposal is arranged.
17. Safety meeting is held with contractor, operations representative, and chemical cleaning coordinator.

VI.4 SAFETY RULES AND EQUIPMENT

As noted in the above checklist, a safety meeting conducted by the chemical cleaning coordinator should precede a chemical cleaning job covering the following points.

1. The contractor's supervisor certifies that he has provided self-contained breathing apparatus and toxic gas detectors in compliance with the Occupational Safety and Health Act, and that all members of his crew are familiar with this equipment, which is kept in an easily accessible location known to the crew.
2. A siren, supplied by the client, is to be sounded by the contractor in the event of an emergency.
3. The coordinator informs the contractor's crew of the client's safety rules covering smoking, specially marked dangerous areas, and other possible local hazards.
4. The contractor agrees to assign at least two men to the job site at all times, equipped with rubber boots, coveralls, gloves, hard hats, safety glasses, and such other safety equipment as is appropriate for the particular job.
5. The contractor is to rope off his work area and post OSHA-approved warning signs to keep unauthorized personnel away.
6. The coordinator and contractor's supervisor review the work order to familiarize themselves with the hookup sketch, the chemical cleaning procedure, any possible hazards, and the safe-

ty precautions to be taken. They verify that:

All blinds are installed as described in the operators' list.

Piping and cleaning connections are installed as specified in the work order.

All drains, manways, vents, steam or condensate lines, or other openings from which acid fumes or other vapors could escape are closed securely.

Chemicals to be used are those specified in the work order.

The work order has been endorsed by the plant manager if hose is to be used instead of pipe.

7. Contractor provides a pressure-tested surge drum connected by screwed or seal-welded piping to a caustic scrubber, flare, or gas recovery system, if hazardous gases may be released.
8. The coordinator assures himself that the contractor's equipment is fitted with adequate controls including pressure gauges, relief valves, thermometers, sample connections, flowmeters, and gauge glasses, and that he has the necessary laboratory apparatus to perform chemical control tests.
9. Before cleaning is started the entire system is hydrostatically tested from the circulation pump, through the unit to be cleaned, to the connection on the surge drum (including the piping from the discharge relief valve to the surge drum), at a pressure at least equal to the shutoff head pressure of the circulation pump.
10. If hose is used for any connections, it should have a minimum working pressure of 500 psi and a bursting pressure of 2000 psi. Hose used for chemical cleaning should contain a neoprene inner-tube covered by three layers of stainless steel braid and a neoprene outer-tube having an antiscuffing surface.*
11. The contractor provides the coordinator with the name, address, and telephone number of his physician and of the hospital where the contractor's crew members are to be taken in case of injury.
12. When hydrogen sulfide is expected to be encountered, sufficient caustic soda must be used in the scrubber to absorb an amount of H_2S equivalent to the total acid to be used in the cleaning process. Rigid seal-welded piping should be used in the presence of H_2S, if possible. (See Chapter VIII, pp. 176-177, for an example of the calculation of sodium hydroxide strength.)

---

* Goodall's Hardrok® Hose is suitable. Goodall Rubber Company, 278 Lawrence Avenue, South San Francisco, California 94080.

VI.5 DISPOSAL OF WASTES

As mentioned previously, the difficulties of waste disposal can be mitigated by minimizing the volume of hazardous waste produced; other possibilities are recovery, recycle, detoxification, and incineration. Acid and alkaline solvents are usually amenable to disposal in existing waste treatment plants if first neutralized and, as a consequence, can often be disposed of on site. It should be realized, however, that such waste disposal in an existing facility must be consistent with the limitations imposed by the National Pollutant Discharge Elimination System (NPDES) permit for that facility. In neutralizing waste solutions it is often expedient to use acidic waste to neutralize alkaline waste provided, of course, that the latter does not contain sulfide. This operation is done by combining the two waste streams in a neutralizing pit in such proportion that the net volume is slightly acidic, then adding a small amount of sodium carbonate to buffer the solution at pH 8.3; adequate mixing is always uncertain in this procedure. Alternatively, waste solutions can be discharged into a lagoon lined with rubber or vinyl, neutralized, and left to evaporate. Acidic cleaning solutions are sometimes neutralized with lime when there is some advantage in precipitating silicate, phosphate, fluoride, ferric, ferrous, and cupric ions. If oil separates on the surface of the neutralized solution, it can be removed by a vacuum truck and burned or else transferred to a recovered oil facility.

Incineration in rotary kilns, which can accommodate a wide variety of wastes, is done in many places, but is precluded by the air quality regulations of some states. O'Neal[10] has reported the results of a study of the effects of spraying an ammonium EDTA solution containing 0.5-1.0 percent iron, 0.2 percent copper, and 0.04 percent nickel into the furnace of a utility boiler. He obtained the best results by injecting the solution below the burners at 58 gpm. At this rate of incineration, approximately 10 percent of the metallic oxides were carried up the stack to give a particulate emission of 0.042 $lb/10^6$ Btu—well below the plant's permit of 0.1 $lb/10^6$ Btu. If the concentration of water in the stack gases exceeded 8.4 percent, it was found that acid smuts were emitted from the exhaust stack, another limitation on the rate of incineration. Solutions of other organic cleaning agents that can be incinerated are citric, formic, and hydroxyacetic acids. The disposal of waste solutions on coal piles[11] or in coal ash pits,[12] as is sometimes recommended, may introduce metallic ions or undesirable organic compounds into runoff water. Other materials that produce air pollutants when oxidized, e.g., sulfamic acid, ammonium bifluoride, should not be incinerated.

Sodium sulfide solutions are often difficult to dispose of, although pulp and paper mills that use the sulfate wood pulping process can make use of them in their black liquor wood digestors. In this process wood chips are cooked in a mixture of sodium sulfate, lime, and water, then, after concentration, the liquor is introduced into a smelting furnace where a hot cinder, called black ash, is formed. The cinder is leached with water to dissolve sodium sulfide and carbonate, after which lime is added to precipitate calcium carbonate; additional sodium sulfate is then added and the liquor is reused.

$$SO_4^{--} + 2C = S^{--} + 2CO_2 \qquad \text{(VI-4)}$$

$$CaO + CO_2 = CaCO_3 \qquad \text{(VI-5)}$$

Trucks used for hauling hazardous liquid wastes must be properly registered and a hazardous waste manifest or bill of lading must be prepared and carried that accurately describes the waste material to be disposed of by either the client or the contractor. The Department of Transportation designates the proper shipping names for various kinds of wastes, their hazard class, and placarding required on trucks. DOT regulations are applicable in all 50 states, including the District of Columbia, Puerto Rico, American Samoa, the Virgin Islands, and Guam; they preempt any inconsistent local regulations. Any other approval and reporting procedures required by public authorities must also be complied with by the client and contractor. Trucking material to approved hazardous waste disposal sites is now the most expensive method of disposal, for the number of dumps is decreasing, and those that remain are becoming more selective of the material that will be accepted.[13]

References

(1) McCoy, J. W. 1969. *Chemical analysis of industrial water.* New York: Chemical Publishing Co.

(2) Am. Soc. Testing Materials. 1980. Recommended practice for sampling and sample preparation techniques in spectrochemical analysis. *Annual Book of ASTM Standards.* Part 42, E311.

(3) Am. Soc. Testing Materials. 1980. Identification of crystalline

compounds in water-formed deposits by x-ray diffraction. *Annual Book of ASTM Standards.* Part 31, E934.

(4) Hughes, R. I., Morgan, T. D. B., and Wilson, R. W. Apr. 19, 1974. Is pyrophoric iron sulfide a possible source of ignition? *Nature* 248: 670.

(5) Am. Soc. Testing Materials. 1980. Recommended practice for measuring coating thickness by magnetic-field or eddy-current (electromagnetic) test methods. *Annual Book of ASTM Standards.* Part 11, E376.

(6) Am. Soc. Testing Materials. 1978. Recommended practice for ultrasonic inspection of metal pipe and tubing. *Annual Book of ASTM Standards.* Part 11, E213.

(7) Am. Soc. Testing Materials. 1980. Recommended practice for radiographic testing. *Annual Book of ASTM Standards.* Part 11, E94.

(8) Stander, R. A. Mar. 17-21, 1958. Precautionary procedures in chemical cleaning, NACE Publication 8A159. Fourteenth Annual Conf., Nat'l. Assoc. of Corrosion Eng., San Francisco, California.

(9) Lawson, M. B. Apr. 6-10, 1981. An economic approach to optimizing a maintenance schedule. Paper No. 77, *Corrosion/81.* The Int'l. Corrosion Forum, Nat'l. Assoc. of Corrosion Eng., Toronto, Ontario.

(10) O'Neal, A. J. Oct., 1977. Experimental incineration of boiler internal cleaning solvent at Long Island Lighting Co. *Combustion* 49 (4): 41.

(11) Hoppe, T. C. and Parmley, J. G. 1972. Preoperational chemical cleaning of utility systems operating at 2000 psi and higher. *Proc. Int'l. Water Conf., Eng. Soc. W. Pa.* 33: 63.

(12) Bell, W. E. and Burguet, J. E. Mar. 14-18, 1977. Considerations in the disposal of spent chemical cleaning solvents. Paper No. 76, *Corrosion/77.* The Int'l. Corrosion Forum, Nat'l. Assoc. of Corrosion Eng., San Francisco, California.

(13) Mallow, A. 1981. *Hazardous waste regulations.* New York: Van Nostrand Reinhold Publishing Co.

VII

Steam Generators

In this chapter several chemical cleaning methods applicable to steam generating plants are considered. These procedures are much more rapid and economical than mechanical cleaning by turbining, hammering, and brushing, all of which, beside being exceptionally disagreeable, require a great deal of labor, if feasible at all. Modern high-pressure boilers cannot be mechanically cleaned; they are designed intentionally to be chemically cleaned. In the United States, the periodic cleaning of boilers with acids and other chemicals is now an accepted routine maintenance practice, although many engineers, especially in Europe, recommend against it. In Germany, for instance, the acid cleaning of a steam generator is a decidedly rare event, whereas in the United States many boilers are cleaned every two or three years. Among the topics covered in the following pages are preoperational cleaning of boiler and preboiler systems, in-service cleaning, and acid and alkaline cleaning of steam generators.

VII.1 PREOPERATIONAL CLEANING

New boilers and auxiliary equipment, particularly that in which heat is exchanged, must be subjected to chemical cleaning before being put in service, in order to remove a variety of possible contaminants including cutting oils, mill scale, grease, lubricating oils, pipe-threading compound, drawing compounds, and protective shop coatings introduced or applied during manufacture and installation of the equipment. The purpose of this preoperational cleaning is to expose the surface of steel so that a normal uniform protective film of magnetite can form when the unit is put in service, and to dissolve or otherwise eliminate particulate matter from the preboiler system that would subsequently be transferred to the steam generator. If this initial cleaning is neglected, pitting and generalized corrosion ensue because covered areas of metal are anodic to clean areas.

Preoperational cleaning should be done as late in the construction schedule as possible, preferably immediately after hydrostatic pressure testing, to avoid the severe rusting and pitting that always occur when aerated water is left stagnant in steel vessels. Hoppe and Parmley(1) recommend that a steam generator be left filled with hydrazine-treated demineralized water under a nitrogen blanket if preoperational cleaning must be delayed after the hydrostatic test. Note, however, that if oxygen is present this may not eliminate pitting.

VII.1.a Preboiler Cleaning

A great many different preboiler arrangements exist, and innumerable combinations of preoperational cleaning methods could be devised. Schlesinger and Burns,(2) for instance, have described nine different chemical cleaning projects in central stations; the authors provide explicit diagrams, and place emphasis upon preliminary planning. The system displayed in Fig. VII.1, in which the dotted lines indicate the flow of solutions through the preboiler system, will suffice here to illustrate the principles of preoperational cleaning. The chemical solution tank, G, and the auxiliary circulating pump, H, may be provided by the client or by the cleaning contractor; in utility power plants permanent cleaning connections, chemical tanks, and pumps for chemical cleaning are common. As shown in Fig. VII.1, the condenser and deaerator are customarily bypassed because mechanical cleaning is preferable for these sizable units in which the volume-to-surface ratio is large. Superheaters and reheaters are also usually bypassed for, if the noise can be tolerated, steam-blowing adequately clears these tubes; air-blowing using soot blower compressors is also a possibility.(2)

The chemical solutions pass through the water side of the low-pressure heaters, through as much of the deaerator piping as possible, then through the high-pressure heaters and economizer, and return to the chemical reserve tank. In many feed water systems, especially those serving boilers operated at less than 900 psi, flushing with hot water at 3-5 fps is sufficient to remove dirt and other siliceous material. Demineralized water, which is often in short supply, is not required for flushing, but the water used should be clear and contain less than 1000 ppm of dissolved solids.(1) Atwood(3) recommends extensive preoperational cleaning for systems operating in excess of 2000 psi, including both tube (water) and shell (steam) sides of feed water heaters. Acidic solvents should not be used in the shells, however, as they are difficult to rinse, and the many baffles they contain trap particulates, preventing effective cleaning.[4]

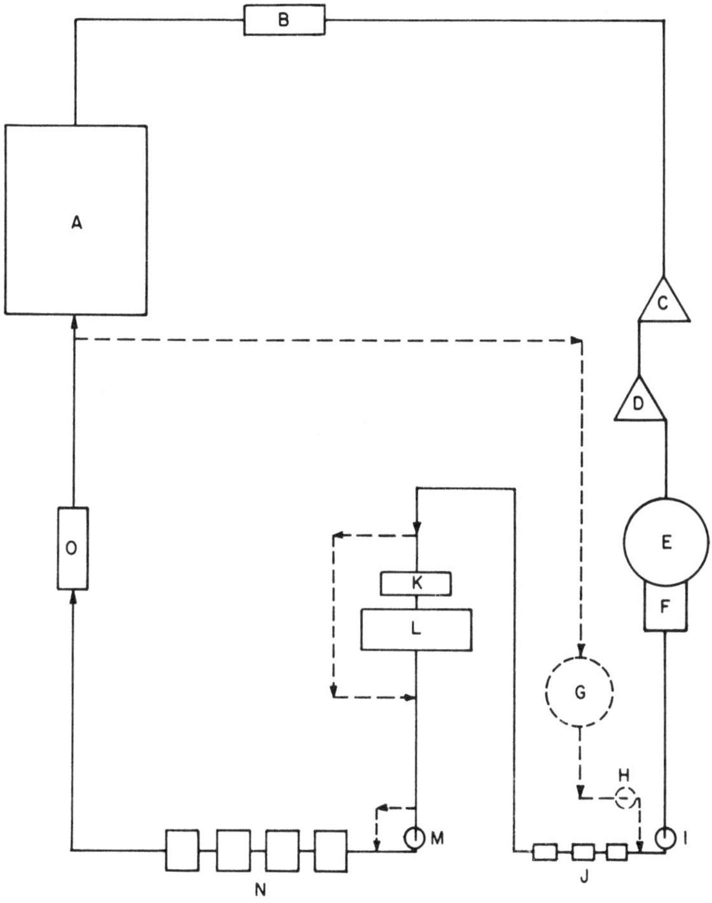

Fig. VII.1

A, Steam generator; B, Superheater; C, High-pressure turbine; D, Low-pressure turbine; E, Condenser; F, Condenser hot well; G, Chemical solution tank; H, Chemical solution pump; I, Condensate pump; J, Low-pressure heaters; K, Deaerator scrubbing section; L, Deaerator storage section; M, Boiler feed pump; N, High-pressure heaters; O, Economizer

Before any aqueous solution is introduced into the preboiler system, miscellaneous trash and construction debris should be removed by hand. Following this, an alkaline solution containing a nonionic wetting agent is prepared in the chemical solution tank, heated to 180-200°F, and circulated through the preboiler system as indicated in Fig. VII.1. In Chapter II, pp. 34-38, several different combinations of alkaline chemicals are described for emulsifying or saponifying oily organic materials. For reasons given there, a 1 percent solution of trisodium phosphate, with the addition of 0.1-0.2 percent of a surfactant, is preferred for routine deoiling. Sodium hydroxide should be avoided, particularly for washing preboiler systems that contain stainless steels, because of the possibility of stress corrosion cracking. For the same reason, water used for preparing alkaline solutions should contain no more than 50 ppm of chloride ion.

If 150-psi steam is available the solution can be heated in a direct contact heater; at lower pressures a heat exchanger is required to prevent excessive dilution. When steam is unavailable it may be necessary to resort to a gas-fired heater, ordinarily one fueled by propane capable of developing 2-3 million Btu per hour. Such heaters are, of course, undesirable in places where hydrocarbons or other flammable chemicals are processed. Circulation of the hot alkaline solution for 6-12 h through the preboiler system, including the shells of the feed water heaters is, in most cases, sufficient. The alkaline solution should remain in the system until it is ready for acid cleaning, as described in Section VII.1.c, at which time the alkaline solution is rinsed out with demineralized water.

VII.1.b Alkaline Boil-Out

Before being put into service all boilers are subjected to a hydrostatic pressure test to check for tightness and strength. The vessel is filled with demineralized and, preferably, deaerated water containing 200-500 ppm of hydrazine, then pressure is applied by means of a pump at one and one-half times the maximum allowable working pressure. At the successful conclusion of the test the water is drained under 5-10 psi of nitrogen until its level in the steam drum, if any, is in the normal operating range. Here it is left until ready for restart or alkaline boil-out if it is a new or retubed boiler.

Practically all new boilers contain oil, grease, and other organic material that must be removed before putting them into service. The undesirable effect of oil or grease in a boiler is to make the surfaces of tubes nonwetting, which in turn leads to film boiling and overheating of the metal. If a contaminating oil is saponifiable, foaming is likely. In addition, oil serves as a nucleus and binder for scaling at hot spots, and it frequently agglomerates boiler sludge and suspended solids into "oil balls," especially in turbulent sections such as steam drums and water wall headers, on account of the roiling of the water in these locations. Also, loose scales containing oil are now and then discovered in water wall tubes. For these reasons, new boilers are deoiled before being put into service or acid cleaned. Before doing this, however, the gauge glass must be removed to prevent etching by the alkaline solvent.

Carbon steel drum boilers operated at low to moderate pressure, e.g., 100-900 psi, can be thoroughly deoiled by circulating a dilute solution containing sodium hydroxide and sodium carbonate; sodium nitrate is sometimes included to prevent stress corrosion cracking. Suppose, for instance, that it is necessary to deoil a newly installed drum boiler having a capacity of 25,000 gal of water. A 50 percent solution of caustic soda along with commercial soda ash is used to prepare the alkaline cleaning solution, which is pumped into the boiler and diluted to 25,000 gal with treated make-up water. The final concentration of each chemical is to be about 3000 ppm. The number of pounds required of each chemical is

$$\begin{aligned} \text{chemical (lb)} &= (3000 \times 25{,}000 \times 8.34)/10^6 \\ &= 625.5 \end{aligned}$$

As 50 percent caustic soda contains 6.365 lb NaOH/gal, the volume needed is

$$625.5/6.365 \cong 98 \text{ gal}$$

The solubility in water of soda ash is 14 percent by weight, at 1.146 specific gravity.[5] Thus, a saturated solution contains

$$(14.0 \times 1.146 \times 8.34)/100 = 1.338 \text{ lb } Na_2CO_3/\text{gal}$$

The required amount of soda ash, 625.5 lb, can therefore be dissolved in a 500-gal mixing tank and then pumped into the boiler. In practice, the boiler is half-filled with treated make-up water and the prepared solution of soda ash is added, after which the concentrated caustic soda is injected and the boiler is filled with water. The pressure is raised to 100 psi and the solution is circulated for 24 h, during which time the concentrations of NaOH and Na_2CO_3 are occasionally checked by the procedure in Chapter XI, p. 236; more chemicals are added, as required, to keep the concen-

water wall headers should be periodically blown down to remove particulates; the loss is made up by adding hot treated water. After dumping the alkaline solution to waste tankage and rinsing once or twice with deaerated fresh water, the unit is ready for restart or acid cleaning. If acid cleaning is not to be done, rinsing may be omitted.

For high-pressure boilers, i.e., 2000 psi, and higher, Hantelman[6] found that a solution containing 7000 ppm Na_3PO_4 and 3000 ppm Na_2HPO_4, along with 0.2 percent of wetting agent was more effective than caustic soda for removing certain tenacious protective coatings. The solution initially contains about 6000 ppm of PO_4^{-3}, but its concentration decreases rapidly during the first hour of heating at 200°F, finally reaching 4000–4500 ppm after 12 h. The phosphate combination has the added advantage of avoiding the hazard of stress corrosion cracking inherent in using sodium hydroxide.

VII.1.c Cleaning with Organic Acids

After completing the alkaline rinse of the preboiler system and the boil-out of the steam generator, the next step in preoperational cleaning is the introduction of an acidic solvent for the purpose of removing mill scale. This step is often omitted in preparing new boilers for operation at less than 900 psi, but it is essential for once-through boilers, as they are not tolerant of particulate matter and do not have provision for blowdown. High-pressure boilers, when contaminated with even small quantities of deposits, are susceptible to high temperature metal failure and chemical corrosion of metal beneath the deposits. Most of the particulate matter that accumulates in a boiler migrates into it from the preboiler system in the form of oxides of iron, copper, and nickel.

The shells of preboiler equipment are usually made of carbon steel, but several special alloys are used for tubing. Thus, condenser tubes may be brass, stainless steel, or titanium; low-pressure heater tubes are either brass or stainless steel; in high-pressure tubes cupro-nickel tubes are common; and economizer tubes are carbon steel, low chromium-molybdenum steel, or stainless steel. Austenitic stainless steels are especially favored in these applications because they are highly resistant to steam, condensate, ammonia, carbon dioxide, and oxygen.[7] They have the disadvantage, however, of being very sensitive to stress corrosion cracking by chloride and, thus, cannot be cleaned with hydrochloric acid. Inhibited organic acids are used instead, the most widely favored being a mixture of hydroxyacetic (glycolic) and formic acids. Citric acid is also appropriate for preboiler systems and once-through boilers that have stainless steel tubes. As all three of these acids are weak, i.e., not highly dissociated, they are easier to inhibit and can be used at higher temperatures than the mineral acids. All of the organic acids must be circulated at a minimum rate of 1 fps to dissolve inorganic deposits, whereas hydrochloric acid performs well in the fill-and-soak method. When cleaning a system that contains no stainless steel, the addition of a small concentration (0.25 percent) of ammonium bifluoride greatly improves the performance of the organic acids, an effect brought about by the stability of the hexafluoroferric ion, FeF_6^{-3}, which has a very small dissociation constant ($K \cong 5 \times 10^{-16}$).

$$Fe_3O_4 + 8H^+ + 12F^- = 2FeF_6^{-3} + Fe^{++} + 4H_2O \qquad \text{(VII-1)}$$

The formation of FeF_6^{-3} has the additional benefit of sequestering ferric ions which, if free in the solution rapidly corrode steel:

$$2Fe^{+3} + Fe = 3Fe^{++} \qquad \text{(VII-2)}$$

Fluoride salts are not recommended in the presence of stainless steels as they too may induce stress corrosion cracking.

The primary purpose of preoperational cleaning is to eliminate mill scale, a mixture of FeO, Fe_2O_3, and Fe_3O_4, formed on steel surfaces at high temperature during fabrication. Although initially deposited as a tight, protective, gray scale, it is brittle and thus tends to flake and erode in service, which creates voids and discontinuities that give rise to electrochemical cells. Except in stressed areas, such as welds, mill scale is very thin—about 0.1-0.5 mil.

A widely used solution for acid cleaning both preboiler systems and steam generators contains 1 percent formic acid, 2 percent hydroxyacetic acid, 0.25-0.5 percent of some proprietary inhibitor, and 0.05 percent of a wetting agent. This solution is heated to 180-200°F and circulated through the preboiler system at 2-3 fps for 4-6 h through the path shown in Fig. VII.1. The inhibitor to be used is normally specified in the contractor's bid, and it must be applied observing the temperature and concentration limits specified by the manufacturer.

As discussed at some length in Chapter II, pp. 61-71 inhibitors are used to impede the attack of hydrogen ions on metal surfaces. The reaction

$$2H^+ + Fe = Fe^{++} + H_2 \qquad \text{(VII-3)}$$

is opposed by an irreversible potential, the hydrogen overvoltage, which is

trations of both near 3000 ppm. During the boil-out the mud drum and markedly affected by the condition of the metallic surface. In addition, the corrosion rate of a metal in an acidic solution is controlled by the rate of reduction of hydrogen ions to hydrogen gas at cathodic sites. This too is influenced by surface properties and by mechanical barriers that interfere with the adsorption of hydrogen atoms, the first step in the evolution of hydrogen gas. Nitrogen bases (triamylamine, pyridine, quinoline, acridine, imidazoline, and numerous others) inhibit the attack of hydrogen ions on steel by raising the hydrogen overvoltage on cathodic surfaces. Inhibition by nitrogen bases is enhanced when they are combined with acetylenic alcohols, such as propargyl alcohol (propyne-1-ol-3), $HC{\equiv}CCH_2OH$, and 2-butyne-1,4-diol, $CH_2OHC{\equiv}CCH_2OH$, which are adsorbed as a film on metallic surfaces. Another combination claimed to be useful[(10)] is thiourea with one of the Mannich bases.[(11)]

Some boiler operators prefer to use the condensate pump to circulate alkaline solutions through the preboiler system, returning the solution to the condenser hot well through temporary piping. This must not be done with acidic solutions, however, and it should also be borne in mind that even in carbon steel systems, valve trim and gaskets are often fabricated from stainless steels and other alloys that are incompatible with certain solvents.

As cleaning progresses samples are taken periodically, preferably every 30 min, and checked for total iron and acidity. The former increases and the latter decreases as time goes by until they eventually stabilize, at which point the cleaning is terminated. If, however, at any time during the cleaning the total acidity becomes less than 1.5 percent, more acid should be added to the circulating solution. As preboiler systems are difficult to drain, the acid solution is usually displaced to waste tankage by demineralized water. This is sometimes followed by a neutralizing solution containing 0.5 percent of sodium carbonate, after which the system is passivated as described in Section VII.1.d.

For reasons given in Chapter II, pp. 56–58, the concentration of inhibited citric acid should be maintained at not less than 3 percent when cleaning a once-through steam generator. When this solvent is used it is circulated at 180-200°F for about 6 h, or until the concentration of total iron becomes constant. Greenberg[(9)] recommends using a 0.5 percent solution of citric acid (pH $\cong$ 2) at 155°F, following the first water rinse, to minimize the precipitation of hydrous ferric oxide; he also advocates nitrogen displacement of the solutions to prevent flash rusting in air.

VII.1.d Passivation

Admiralty brass, bronze, cupro-nickel, monel, Muntz metal, and stainless steels are all self-passivating in aqueous solutions. Carbon steel after being cleaned with acid, however, is left in a highly reactive state in which the surface rusts rapidly upon exposure to air. In order to passivate steel it is, therefore, necessary to apply one of several possible solutions listed in Table V-1 , p. 109; many other details of the procedure are supplied in Chapter V, pp. 104–112, that are not repeated here.

At the conclusion of a chemical cleaning, acid solutions are displaced or drained under a blanket of nitrogen to exclude air. It is essential to keep the inside of a boiler as free from air as possible to avoid extensive oxygen pitting along mandrel or draw markings. To the unaided eye these appear to be cracks or grooves, but when magnified, they are seen to be strings of pits. Several rinses may be required (as indicated by pH) to remove all traces of free acid; for this reason some cleaning contractors also recommend a neutralizing step. As mentioned in Section VII.1.c, it is worth taking the trouble to include a rinse with dilute acid (pH $\cong$ 2) to solubilize hydrous ferric oxide, which precipitates by hydrolysis at pH 3, or above.

$$Fe^{+3} + 3H_2O = Fe(OH)_3 + 3H^+ \qquad \text{(VII-4)}$$

When the pH of the rinse water indicates that acid is completely flushed out, a passivating solution is heated, pumped into the equipment to be passivated, and circulated for 1 h at elevated temperature. Alternatively, when passivating a boiler, it can be lightly fired to about 100 psi, then cooled and drained. This treatment yields a passive surface on acid-cleaned steel that resists rusting when the equipment is subsequently opened for inspection and repairs. Duncan, et al.[(8)] have found that the following recipe, prepared with demineralized water or condensate, provides even, effective passivation in a short period of time.

0.25 percent NaH_2PO_4
0.25 percent Na_2HPO_4
0.25 percent $NaNO_2$

In use, the solution is heated to 200°F, circulated for 15 min, allowed to soak for 30 min, then drained. Steel surfaces so treated remain rust-free for at least 3 days in a moist atmosphere. Note that if the above solution is exposed to an excess of acid, the very toxic gas, nitrogen dioxide, is evolved, as evidenced by brown fumes of NO_2. Even at low concentrations this gas rapidly causes severe damage to the lungs.

$$2H^+ + 2NO_2^- = NO + NO_2 + H_2O \qquad \text{(VII-5)}$$

For other methods of passivation refer to Chapter V and to Section VII.3, below.

VII.2 IN-SERVICE BOILER CLEANING

As in-service cleaning is in most instances carried out over a relatively long period of time by boiler plant personnel rather than by chemical cleaning contractors, the subject is in some measure outside the scope of this work. Boilers are often cleaned this way, however, so the methods deserve a few words, at least.

Experienced boiler operators are aware that relatively minor changes in the composition of make-up water, alterations in internal treatment, or radical changes in blowdown rate can cause scale to slough or spall from a scaled boiler, blocking tubes and headers, plugging blowdown lines, and obstructing circulation. For this reason it is advisable to be conservative when changing from one internal treatment to another. A related concept is that of removing scales by deliberately altering chemical treatment, a process that has become known as in-service cleaning. It is not always possible to predict the effect of a chemical on a scale, but, in general, phosphate attacks calcium carbonate rapidly, causing foaming (effervescence, actually, as much carbon dioxide is released) and high alkalinity.

$$3PO_4^{-3} + 5CaCO_3 + 5H_2O = Ca_5(OH)(PO_4)_3 + 10OH^- + 5CO_2 \qquad \text{(VII-5)}$$

In the presence of carbonate, phosphate also transforms calcium sulfate very rapidly.

$$CO_3^{--} + CaSO_4 = CaCO_3 + SO_4^{--} \qquad \text{(VII-6)}$$

This reaction is followed by that shown in Eq. (VII-5). If the intention is to remove calcium sulfate alone, however, it is preferable to use sodium hydroxide, as it is then easier to control alkalinity and the foaming caused by the evolution of carbon dioxide is avoided.

$$3PO_4^{-3} + 5CaSO_4 + OH^- = Ca_5(OH)(PO_4)_3 + 5SO_4^{--} \qquad \text{(VII-7)}$$

Calcium silicate is slowly dissolved by phosphate, but elevated alkalinity more rapidly attacks this and other silicates; the addition of silicates themselves is effective for fluidizing magnesium sludges.

$$CaSiO_3 + 2OH^- = Ca^{++} + SiO_4^{-4} + H_2O \qquad \text{(VII-8)}$$

$$Mg_3(PO_4)_2 \cdot Mg(OH)_2 + 4SiO_3^{--} = 4MgSiO_3 + 2PO_4^{-3} + 2OH^- \qquad \text{(VII-9)}$$

It is unlikely that any of these reactions go to completion in a boiler; their major effect is to loosen the connection between scale and metal. When contemplating a change in treatment in an unusually dirty boiler containing voluminous deposits, extensive scale, or copious amounts of iron oxides and copper, it is advisable to acid-clean first, as described in Section VII.3. This precaution applies especially to waste heat boilers, which are prone to develop heavy deposits of magnetite, Fe_3O_4, and hematite, Fe_2O_3, in their shells and on their tubes. As the latter are nearly always triangularly pitched, they cannot be effectively cleaned by water jetting. Vertical waste heat boilers should be chemically cleaned more frequently than they usually are, for when badly fouled boilers are eventually acidified, copious amounts of scale fall onto the lower support plates, initiating the phenomenon of denting, or deformation, of the tubes where they pass through the plate. Pettit and McTigue(12) have reported a similar effect on Iconel-600 tubes supported by carbon steel plates in nuclear power plants. Denting is caused by the growth of nonprotective magnetite in the crevices between the tubes and the tube support plates. Chelants should never be used in attempting an in-service cleanup of waste heat boilers, for regardless of claims to the contrary, when severely fouled, they cannot be recovered by altering the water treatment.

Herman and Gelosa(13) have alluded to an internal treatment for sludge removal wherein polymers, "organics," sequestrants, and iron sequestrant-dispersant are added at unspecified concentrations to accomplish a slow, on-stream cleanup. Lignins or carboxymethylcellulose are used in their usual capacity as protective colloids to disperse sludge; to supplement the dispersants, a few parts per million of chelant are included, which seems to assist in loosening existing deposits; hydroxyethylidenediphosphonate, or other phosphonate, may be included to assist in disintegrating deposits coagulated by iron. Approximately three months are required to dissipate sludges and deposits that are at all susceptible to the treatment.

Edwards and Rozas(14) have reported the pioneering work of the Dow Chemical Company, starting in 1959, in continuous internal treatment of boilers with EDTA. Subsequently,[15] the possibilities of this treatment for in-service cleaning of drum boilers became apparent. Although chelants cannot sequester iron oxides, they do chelate calcium and magnesium in deposits and sludge in which iron functions as a binder. This releases particles of iron oxide, which are then removed in the blowdown. Thus, within a few hours of initiating a chelant treatment in a dirty drum boiler, quantities of suspended iron oxide appear in the blowdown water. Also, when a phosphate program is replaced by chelant, the concentration of phosphate ion in the boiler water is likely to be higher for several days than it was before discontinuing the addition of phosphate, because of some dissolution of calcium and magnesium phosphates. When using chelants for in-service cleaning, the boiler operator should be aware of the corrosiveness of excessive concentrations of these chemicals in boiler water.[16]

VII.3 OPERATIONAL CLEANING OF DRUM BOILERS

Deposits in moderate-pressure boilers usually are composed of hematite, magnetite, silicates, and alkaline earth salts; at pressures above 2000 psi laminar scales of hematite, magnetite, various spinels, and metallic copper are common. The location and thickness of copper scales can be ascertained by x-ray radiography, but if other deposits are present, Wackenhuth and Richards[17] recommend removing a water wall tube to determine whether chemical cleaning is necessary and, if so, to find an effective solvent for the purpose. The choice of cleaning agent is to some extent empirical, so some sort of performance tests in a laboratory may be desirable to assess solubility in various chemicals, to find the most suitable temperature, and to estimate the efficiency to be expected.[4] In this section three more-or-less standard procedures are given for cleaning boilers with and without copper-containing deposits.

VII.3.a Routine Acid-Cleaning

Operational acid-cleaning may be undertaken for any of several reasons:

1. to reduce the thickness of the magnetite film and to dissolve insulating deposits of copper, zinc, nickel, and iron oxides;
2. to prepare the boiler for a change in internal treatment;
3. to achieve an immediate improvement in heat transfer;
4. to avoid troublesome unscheduled shutdowns caused by tube failures.

Rice[18] has reported, however, that repeated acid cleanings can damage areas with heavy local stress, such as field welds, improperly rolled tubes, cold-worked metal, bolt threads, and draw marks on tubes. Longitudinal cracking, roughening of tube ends, deep etching, pitting, and honeycombing have all been seen after acid cleaning, the severity of attack being influenced to a great extent by the kind of deposit in contact with the boiler metal when cleaning commences. Thus, ferric ions liberated in the acid dissolution of hematite and magnetite are highly corrosive to boiler steel because of the large positive oxidation potential of the reaction represented by Eq. (VII-12).

$$Fe_2O_3 + 6H^+ = 2Fe^{+3} + 3H_2O \qquad \text{(VII-10)}$$

$$Fe_3O_4 + 8H^+ = 2Fe^{+3} + Fe^{++} + 4H_2O \qquad \text{(VII-11)}$$

$$2Fe^{+3} + Fe = 3Fe^{++} \qquad \text{(VII-12)}$$

From a consideration of molecular weights in these reactions it can be seen that 160 lb of Fe_2O_3 or 232 lb of Fe_3O_4 solubilize 56 lb of boiler steel as ferrous ion, and also that these amounts represent one-third and one-fourth, respectively, of the total iron brought into solution by acids. Ferric ion corrosion, as it is customarily called, is mentioned briefly in Chapter II, pp. 73–75, but it can cause such severe damage in boilers undergoing acid cleaning that a few of the methods for masking or eliminating ferric ion deserve comment here.

Ferric ions can be reduced to ferrous ions, or they can be sequestered, thereby reducing the oxidation potential of Eq. (VII-12). In addition, special filming inhibitors can be applied that shield steel surfaces. Engle and Keeney[19] have disclosed a method for reducing ferric ions with H_2S produced in a cleaning solution by the hydrolysis of thioacetamide, CH_3CSNH_2. Their formulation also includes an acetylenic alcohol, an amine, and formaldehyde. The first two are conventional inhibitors of acid corrosion. Thioacetamide, when heated, hydrolyzes to acetamide, CH_3CONH_2, and H_2S, which in turn reduces any ferric ions, while any excess H_2S reacts with formaldehyde to form the insoluble heterocyclic compound, trithiane. The sequence of reactions, the last three of which are quite rapid, is as follows.

$$Fe_2O_3 + 6H^+ = 2Fe^{+3} + 3H_2O \qquad \text{(VII-13)}$$

$$CH_3CSNH_2 + H_2O = H_2S + CH_3CONH_2 \qquad \text{(VII-14)}$$

$$2Fe^{+3} + H_2S = 2Fe^{++} + S + 2H^{+} \qquad \text{(VII-15)}$$

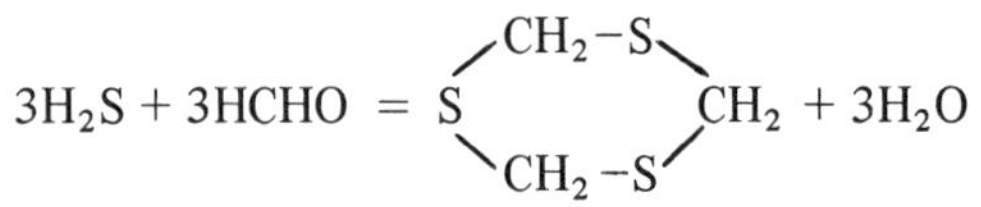

(VII-16)

Because of the possible formation of s-dichloromethyl ether, hydrochloric acid cannot be used with formaldehyde in a chemical cleaning solution; dilute sulfuric acid should be used instead.

$$2HCHO + 2HCl = ClCH_2{-}O{-}CH_2Cl + H_2O \qquad \text{(VII-17)}$$

This dangerous chlorinated compound, if inhaled, can cause oat cell carcinoma, a form of lung cancer.[20]

In principle, any chemical having an oxidation potential less negative than -0.77 V can reduce ferric to ferrous ion. Pettit and McTigue,[12] reporting on a joint project of the U.S. Department of Energy and the Consolidated Edison company of New York to develop methods for cleaning steam generators in nuclear power plants, mention a combination of citric and ascorbic acids as a prospective cleaning solvent. Although the authors make no comment on the function of ascorbic acid, it must inevitably reduce ferric ions while being itself oxidized to dehydroascorbic acid.

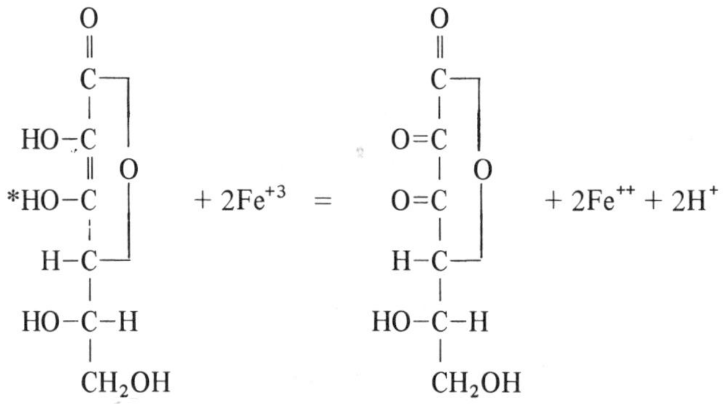

(VII-18)

Isoascorbic acid, also called among other things, erythorbic acid, has recently been proposed as a substitute for hydrazine as an oxygen scavenger in boilers.[21] Its reaction rate with oxygen is rather slow, but ferric ion in acidic solutions is quickly reduced by either isomer. The hydrogen atom marked with an asterisk in Eq. (VII-18) is easily replaceable by a sodium ion in both acids to form soluble sodium salts.[22]

Conventional amine corrosion inhibitors, although capable of restraining attack by acids, are totally ineffectual in controlling ferric ion corrosion. Another group of compounds, however, the benzylic sulfonium salts, are reduced on steel surfaces to a stable hydrophobic film that stifles both anodic and cathodic corrosion reactions, while simultaneously serving as a physical barrier against ferric ions. Sulfonium salts have an unshared electron pair and, thus, are capable of forming coordination complexes with the transition elements.[23] Frenier and Settineri[24] have disclosed the application of particular benzylic sulfonium salts as inhibitors of ferric ion corrosion; more recently, they have elucidated the mechanism of inhibition.[25] When dodecylbenzyldimethylsulfonium chloride (I), for example, comes in contact with iron it is reduced to a surface film of p,p′-bisdodecylbibenzyl (II), p-dodecyltoluene (III), and dimethylsulfide (IV). In the following equations $R = C_{12}H_{25}C_6H_4CH_2-$.

$$\underset{\text{I}}{2R\!-\!\overset{+}{S}\!-\!(CH_3)_2Cl^-} + Fe = \underset{\text{II}}{R\!-\!R} + \underset{\text{IV}}{2(CH_3)_2S} + Fe^{++} + 2Cl^- \qquad \text{(VII-19)}$$

$$\underset{\text{I}}{2R\!-\!\overset{+}{S}\!-\!(CH_3)_2Cl^-} + 2Fe + 2H^+ = \underset{\text{III}}{2R\!-\!H} + 2(CH_3)_2S + 2Fe^{++} + 2Cl^- \qquad \text{(VII-20)}$$

The sulfonium cation itself is not detectable in the hydrophobic protective film.

Another complication that is often mentioned, but seldom happens, is the entrapment of acid cleaning solution beneath deposits that chemical cleaning has failed to remove. When the boiler is subsequently fired, serious hydrogen damage occurs in these locations. The possibility of acid remaining in a boiler after proper chemical cleaning is remote, in any event, and can be eliminated completely by water jetting the tubes to remove the insoluble debris that remains after rinsing and passivation; this also makes the subsequent inspection for pitting and cracking much easier. Although there is always some possibility of this type of damage being done by acid cleaning, in most instances extensive pitting within a boiler is the result of using aerated water for rinsing, omitting nitrogen blanketing during drainings, or leaving the wet boiler exposed to air for an extended period after cleaning. Atwood[3] points out that the most likely cause of damage when acid-cleaning boilers is the existence of localized high metal temperatures within the boiler when the acid solution is introduced. Regardless of all of this, it is prudent to check the soundness of tubes and the extent of corrosion by ultrasonic procedures following an acid cleaning.

The solvent most often chosen to clean drum boilers is a dilute solution of hydrochloric acid to which an inhibitor is added to minimize attack on the metal being cleaned. Most water-formed deposits, including phosphate sludges, alkaline earth salts, thickened layers of magnetite, and silicate scales dissolve readily in inhibited dilute hydrochloric acid, although if alkaline earths and silicates are present in concentrations exceeding 3-5 percent, the fouling may be difficult to dissolve. The addition of ammonium bifluoride significantly increases the rate of dissolution of both silicates and magnetite. When the latter is detached from metal it dissolves very slowly in hydrochloric acid because reductive dissolution, represented by Eq. (VII-21), plays no role in the solution process.

$$Fe_3O_4 + Fe + 8H^+ = 4Fe^{++} + 4H_2O \qquad \text{(VII-21)}$$

A satisfactory cleaning solution for drum boilers that do not contain stainless steel components or deposits of copper is prepared as follows:

5.0–7.5	percent hydrochloric acid*
0.25–0.75	percent ammonium bifluoride
0.2–0.3	percent proprietary inhibitor
0.03	percent nonionic wetting agent, e.g., an alkylarylpolyethoxy alcohol

Stander[26] considers the maximum allowable corrosion rate in cleaning a boiler with inhibited hydrochloric acid to be 0.8 $mg/cm^2 \cdot h$ (0.04 mil/h). The corrosivity to carbon steel of the stronger of the above mixtures was found to be 260–350 mpy (0.03–0.04 mph), with no evidence of pitting.

Proprietary inhibitors must be used at the concentrations and temperatures prescribed by their manufacturers. They should not be added to inventory acid, although some contractors do this, for it is unwise to rely on the shelf-life of inhibitors in concentrated acids, which induce polymer-

*Some boiler manufacturers limit the concentration of hydrochloric acid to 5 percent.

ization in some products. The proper volume of inhibitor is added to the concentrated acid shortly before cleaning is to start. The strengths of hydrochloric acid and ammonium bifluoride to be used depend upon the history of the boiler to be cleaned. In general, if a boiler has not been cleaned for several years, or if inspection reveals extensive deposits, a solution containing 7.5 percent of hydrochloric acid and 0.75 percent of ammonium bifluoride is appropriate. On the other hand, if the boiler has been cleaned within the preceding two years, if the water treatment is known to be satisfactory and no upsets have occurred, and if only light deposits are found on inspection, then the lesser strengths of 5.0 percent hydrochloric acid and 0.25 percent ammonium bifluoride are sufficient. Atwood(4) has demonstrated that hydrochloric acid is the only cleaner that dissolves scale in a reasonable length of time under the static condition that exists when a boiler is filled and soaked without circulation; all other common cleaning agents must be circulated at at least 1 fps to remove scale completely at a reasonable rate.

If organic sludge conditioners such as lignins, tannins, or quebracho have been used for internal treatment, or if the presence of oil is suspected, an alkaline boil-out should be done before attempting to acid clean. This is accomplished by firing the boiler at about 100 psi while injecting the alkaline solutions described in Section VII.1.b , pp. 134-136, through the chemical injection pump(s), then blowing down the boiler until the pH of the water is reduced to about 10. It is now the consensus in the United States that all new boilers operating at pressures above 900 psi should also be acid cleaned after alkaline boil-out.

In the normal course of events, a boiler is scheduled for chemical cleaning at one of the times when it is brought down for recertification. The cleanliness of the internal surfaces is improved by doubling the normal blowdown rate during the two weeks preceding shutdown, while at the same time increasing the dosage of sludge conditioner to 1½ times normal. Also, the boiler should not be drained until it is relatively cool, i.e., less than 212°F. Preparatory to acid cleaning, the tubes are flushed with hot boiler feed water to remove any loose debris. When the wash is completed, vents are opened, draft fans are started, and the boiler is cooled to 160–180°F, an attempt being made to achieve this temperature range by the time the cleaning contractor is scheduled to arrive. All furnace openings are closed, including drafts, dampers, and burner doors, to retain heat in the boiler; the latter must never be heated by direct firing.

Boilermakers next isolate all lines and equipment that should not be exposed to hydrochloric acid (stainless steel, in particular), then remove the blowdown valves on the water wall headers, replacing them with 2-in. screwed connections. A drain is installed on the lower drum and a 2-in. screwed connection is also installed either on the steam drum just above the normal water level, or else on the upper arm of the level controller. After repairing any leaks that may have developed, the boiler is turned over to the chemical cleaning contractor, who connects suitable hoses from the discharge manifold of his chemical solution pump to each of the blowdown nozzles provided; another hose connects the steam drum to the solution tank from which the chemical pump takes suction.

Sometime before filling the boiler with inhibited acid solution, the superheater is back-flushed with water until it is filled and overflows into the steam drum. This is done to prevent the inadvertent entry of acid should the steam drum be overfilled, and to prevent exposure to acidic vapor should the temperature of the boiler be so high that the acid solution boils. Neither eventuality is likely if an experienced contractor is employed, but the precaution is worthwhile.

Before pumping the cleaning solution into the boiler, it is the responsibility of the contractor to verify that the temperature of the boiler is in the specified range. This is often done by measuring the temperatures at the ends of the boiler drums, although this is inexact, as the ends of the drums are the thinnest accessible parts of the boiler and, thus, do not reflect what the temperature might be in the interior, where there is a great mass of metal not directly exposed to the cooling fans. Because of this, the contractor may first fill the boiler with tempering water, externally heated to a few degrees below the desired temperature if the readings at the drum ends are too high, or a few degrees above the desired temperature if the readings are too low. If, after passing through the boiler, the temperature of the tempering water is in the correct range, the boiler may be filled with acid solution, pumped in at a little above or below the desired temperature, according to circumstances. As it is generally agreed that no more than 1 h should be needed to fill a boiler with acid, the contractor is expected to provide enough pumping capacity to accomplish this, and also to provide a source of heat adequate to maintain the acid solution in the desired temperature range throughout the cleaning.

Because of the large volume of cleaning solution required, a flow of water is started into the boiler and a 37 percent solution of hydrochloric acid, verified to be suitably inhibited, is blended into the water to give the desired concentration. This requires frequent titrations of acid strength and adjustments of the flow of acid during the filling. The boiler is filled through the lower blowdown connections until the solution overflows into the tank truck or chemical solution tank through the hose connected to the steam drum. The usual procedure is to allow the boiler to soak without circulation for 4-6 h at 160°F, which is sufficient time for hydrochloric acid to dissolve most water-formed deposits. It is better practice, however, to use the filling pump to maintain a slow circulation throughout the entire period. The pump augments thermal circulation, continuously delivering fresh acid to the internal surfaces. Even though the total volume is much greater than the capacity of the inlet nozzles, their throughput is still enough to provide slow circulation and also to reduce the temperature of any hot spots that might exist. In some instances, when scale is exceptionally heavy, a longer treatment might be needed, but this can only be determined if the boiler has first been opened for inspection. As circulation is slow and the volume of acid is large compared to the weight of scale contacted, there is no need to measure the concentration of acid during the soaking period, particularly if there is no means for obtaining a representative sample, as in the fill-and-soak procedure. In the event that a positive circulation rate of at least 1 fps can be maintained, it is desirable to draw a sample of the cleaning solution every 30 min and determine the acidity and concentration of total iron by the procedures in Chapter XI. If the concentration of hydrochloric acid falls below 2 percent, drain the solution under nitrogen, and refill with a fresh solution of 5 percent hydrochloric acid. There is not much point in attempting to measure ferric ion in cleaning solutions for, as mentioned on p. 75, there is little likelihood of finding any in a solution that has been in contact with steel tubes, boilerplate, pumps, and sample lines. For those who think that the concentration of ferric ion in a cleaning solution is significant, a method is provided in Chapter XI for its determination.

When cleaning is judged to be complete, as indicated by the acidity and total iron results, the acid solution is forced out through the lower drain hose under 5-10 psi of nitrogen into waste tankage or a vacuum truck for disposal. All drainings and refillings of the boiler should be done under nitrogen blanketing, and as quickly as possible. After the acid has been drained, the boiler is filled with hot, deaerated boiler feed water, then drained. The boiler is next rinsed for about 30 min with a 0.5 percent solution of citric acid or hydrochloric acid—pH about 2—at 160°F, to prevent the precipitation of $Fe(OH)_3$, after which the solution is drained. Finally, the boiler is passivated with a nitrite-phosphate solution as described in Section VII.1.d, above. After draining the passivating solution, the boiler may be opened for inspection; before entering, the boiler should be checked for the presence of explosive gases.

At this point a large volume of loose rubble frequently remains in headers, the mud drum, and in horizontal tube runs; this can be flushed out by water jetting at 6000–7000 psi using rubber-lined tubing and boiler lances. After inspecting, repairing drum internals, grinding out any grooves in the drum shells or tube ends, and doing any other necessary maintenance, the boiler is reclosed and hydrotested, using treated water containing hydrazine or sulfite, as outlined in Section VII.1.b.

VII.3.b Acidic Copper Removal

As scales and deposits in boilers operated at pressures above 1250 psi are, in the main, products of preboiler corrosion, they typically contain iron, copper, zinc, nickel, and sometimes aluminum. According to Noll,[27] products of corrosion in the preboiler system are the largest single cause of tube failures in high-pressure boilers. Corrosion products include copper oxides, ferric oxide, magnetite, nickelous oxide, zinc oxide, and occasionally, chromic oxide, all of which are introduced as particulates into feed water by the corrosion of ferrous and nonferrous alloys in the condensate and preboiler systems. Metallic copper is often present and spinels have also been identified by x-ray diffraction.[17] The latter are very hard aluminates or ferrates of magnesium, zinc, copper, or nickel, e.g., $CuAl_2O_4$. These deposits are usually found on the inside surfaces of boiler tubes on the sides exposed to the burners.

Condenser tubes are fabricated of admiralty brass, aluminum brass, and stainless steels. Preheaters may be made of admiralty brass, stainless steels, or in high temperature stages, cupro-nickel, although the latter is subject to exfoliation on the steam side and to the elaboration of copper foils on the water side.[28] Economizers, shells, piping, and so on are usually carbon steel and, thus, a source of particulate magnetite in boilers. Corrosion products from the alloys attach themselves to boiler tubes at places where heat flux is high, where the flow of water is relatively slow, just beyond bends in boiler tubes that change direction from vertical to horizontal, and at rough spots such as butt-welded tube joints. At pressures above 850 psi, deposition is virtually complete and these sintered deposits from a porous, insulating layer that also serves as a trap for dissolved solids, which when concentrated cause pitting and caustic gouging. Neat[28] and Long[29] have both recommended the replacement of nonferrous alloys in preboiler

At pH 9–10 there is a substantial excess of NH_4OH in the solution favoring the formation of the cupric ammonia complex, $Cu(NH_3)_4^{++}$, when copper is oxidized by sodium nitrite.

$$2NO_2^- + 2Cu + 2H_2O + 8NH_3 = N_2O_2^{--} + 2Cu(NH_3)_4^{++} + 4OH^- \quad \text{(VII-23)}$$

Water is circulated in the boiler, sufficient citric acid is added to give a concentration of 3 percent, and the pH of the solution is adjusted to 3.5–4.0. The circulation rate must be at a minimum of 2 fps for 6–8 h, more citric acid being added, as necessary, to maintain its concentration at 3 percent. The solution is then cooled to 140–150°F and its pH is raised to 9.5–10.0 by injecting ammonia. When the pH is in the proper range, sodium nitrite is added to give a concentration of 0.5–1.0 percent, after which the solution is reheated to 180–210°F and circulated for 2–4 h. Some contractors agitate the solution with air admitted through lower blowdown connections. Finally, the solution is drained, and the boiler is rinsed with hot treated boiler feed water or condensate. A separate passivation step is unnecessary after this treatment.

An efficient method for removing copper from high-pressure boilers with an alkaline solution of EDTA is outlined in Chapter II, pp. 81–83. Special advantages of the procedure are very low corrosion rates (0.3–0.5 mpd) and easy waste disposal.

VII.4 TURBINES

Although not usually considered so, a steam turbine is really part of the steam system and, thus, is subject to fouling by salts and other particulates carried through the superheater in steam. Howell and McConomy,[32] for instance, using x-ray diffraction, have identified deposits of nepheline, $K_2Na_4Al_6(SiO_4)_6$, and noselite, $4Na_2O \cdot 3Al_2O_3 \cdot 6SiO_2 \cdot SO_4$, in the early, high-pressure stages of a 2075-psi turbine and α-quartz and magnetite in the low-pressure stages. Silicic acid that volatilizes with steam, if in excess of 0.02 ppm SiO_2, ultimately deposits in turbines, reducing their efficiency and, in instances of severe contamination, unbalancing the rotor. Straub and Grabowski[33] found that deposits in turbines driven by steam at 600 psi were composed of sodium chloride, sodium sulfate, sodium hydroxide, and sodium silicate, and were, in general water-soluble. At higher pressures, however, three forms of silica are deposited on turbine blades. At the high-pressure end, where temperatures are in the range of 500-700°F, the predominant structure is sodium disilicate, $Na_6Si_2O_7$, the sodium salt of the ortho form of disilicic acid. This salt dissolves in water and is strongly hydrolyzed, so that it titrates as sodium hydroxide, although x-ray diffraction shows that the latter is not actually present in the deposit.

$$Na_6Si_2O_7 + 3H_2O = 6Na^+ + 6OH^- + 2SiO_2 \qquad \text{(VII-24)}$$

Further along in the turbine, where the temperature falls into the range of 300-500°F, less soluble forms of silica separate, including α-quartz, cristobalite, and chalcedony. These sometimes yield to caustic washing. Finally, at the low-pressure end, where the temperature is 100-300°F, extremely insoluble amorphous silica is deposited, which in most cases must be removed by mechanical cleaning, e.g., by alundum blasting. These silicates separate as they do because, although they enter the turbine dissolved in steam, as pressure and temperature fall in the course of its passage through the machine, the steam becomes supersaturated with respect to silicic acid. At the hot end there is time for crystalline structures to grow, but toward the cold end the temperature is falling so rapidly and the rate of precipitation is so fast that silica solidifies before it has time to crystallize. Amorphous forms of most inorganic chemical compounds are very slightly soluble in water.

Although high steaming rates sharply increase the concentration of silica in steam, much more silica is transferred to the turbine during boiler restarts than in continuous operation. Then too, oxides formed by corrosion are at elevated concentrations in the condensate and feed water during start-ups; among these, copper is particularly troublesome when deposited on throttle valve seats or in turbines. At ultra-high pressures sodium salts and copper oxides appear to steam-distill in a manner similar to silica, for they too pass through the superheater without any significant deposition there. During pressure reduction from 3500 to 2400 psi, however, cuprous and cupric oxides precipitate from the steam in the high-pressure stages of the turbine, seriously impairing its performance.[34] Morris and Call[35] have reported their experience in flooding turbines with a patented copper solvent[36] containing 18 percent ethylenediamine and 7.5 percent of hydroxylamine hydrosulfate. Both cuprous and cupric ions form stable coordination complexes with ethylenediamine.

$$Cu^+ + 2NH_2CH_2CH_2NH_2 = Cu[NH_2CH_2CH_2NH_2]_2{}^+ \qquad K = 6.3 \times 10^{10} \qquad \text{(VII-25)}$$

$$Cu^{++} + 2NH_2CH_2CH_2NH_2 = Cu[NH_2CH_2CH_2NH_2]_2^{++} \qquad K = 1.0 \times 10^{20} \tag{VII-26}$$

The role of hydroxylamine in the solvent is uncertain, but it readily decomposes to ammonium ion and nitrous oxide, as indicated in Eq. (VII-27); the reaction is favored in alkaline solution.

$$4NH_3OH^+ = N_2O + 2NH_4^+ + 3H_2O + 2H^+ \tag{VII-27}$$

$$4NH_2OH = N_2O + 2NH_3 + 3H_2O \tag{VII-28}$$

As it is always worthwhile to attempt to remove turbine fouling by washing, the machines should be permanently piped to a supply of hot treated water. The turbine to be washed must be cool before starting the process. Water is admitted to the wash line and the drain valve is opened to warm up the line. The throttle valve to the turbine is gradually opened until the machine has settled at a speed of 300–350 rpm. If possible, the turbine should be started by admitting steam through the bypass around the stop valve to protect against excessive speed. Finally, the valve in the wash water line is opened to admit water until a spray of steam and water issues from the drain in the exhaust line from the turbine. If the fouling is susceptible to water, 3 h is sufficient time to wash a turbine. The determination of sulfate is sometimes used to follow the washing of turbine blades when freeing them from deposits consisting primarily of sodium hydroxide and sodium chloride, with lesser amounts of sodium sulfate and sodium silicate. During the washing, samples are taken periodically and sulfate is determined. Ordinarily, sodium hydroxide and sodium chloride are washed off of the blades faster than sulfate, so the latter is used as an indicator of the completeness of the washing.

References

(1) Hoppe, T. C. and Parmley, J. G. 1972. Preoperational chemical cleaning of utility systems at 2000 psi and higher. *Proc. Int'l. Water Conf., Eng. Soc. W. Pa.* 33: 63.

(2) Schlesinger, H. A. and Burns, V. T. 1972. Planning and application preoperational cleaning of fossil fueled central station units. *Proc. Int'l. Water Conf., Eng. Soc. W. Pa.* 33: 81.

(3) Atwood, K. L. 1972. A review of pre-operational cleaning practices in high pressure utility systems. *Proc. Int'l. Water Conf., Eng. Soc. W. Pa.* 33: 73.

(4) Atwood, K. L. Sept., 1970. Solvent selection for preoperational and operational cleaning of utility boilers. *Combustion* 42 (3): 16.
(5) Dean, J. A. 1973. *Lange's handbook of chemistry.* 11th ed. New York: McGraw-Hill.
(6) Hantelman, W. D. 1972. Ten years experience in preoperational chemical cleaning of power plant steam generating systems. *Proc. Int'l. Water Conf., Eng. Soc. W. Pa.* 33: 101.
(7) Efird, K. D. and Moller, G. E. Mar. 6–10, 1978. Electrochemical characteristics of 304 and 316 stainless steels in fresh water as functions of chloride concentration and temperature. Paper No. 87. *Corrosion*/78. The Int'l. Corrosion Forum, Nat'l. Assoc. of Corrosion Eng. Houston, Texas.
(8) Duncan, F. R., Fyock, D. H., Baldwin, L. V., and Hopkins, R. D. 1972. Preoperational cleaning of high pressure boilers of General Public Utilities Corporation. *Proc. Int'l. Water Conf., Eng. Soc. W. Pa.* 33: 97.
(9) Greenberg, S. 1966. Factors that must be considered for successful chemical cleaning as experienced in naval boilers. *Proc. Am. Power Conf.* 28: 818.
(10) Anderson, J. D., Hayman, E. S., Jr., and Rodzewich, E. A. Nov. 16, 1976. *Acid inhibitor composition and process in hydrofluoric acid chemical cleaning.* U.S. Patent No. 3,992,313.
(11) Mannich, C. and Krösche, W. 1912. Ueber ein kondensationsprodukt aus formaldehyd, ammoniak und antipyrin. *Arch. Pharm.* 250: 647.
(12) Pettit, P. J. and McTigue, P. F. 1978. Chemical cleaning of PWR steam generators to relieve denting. *Materials Performance* 17 (10): 36.
(13) Herman, K. W. and Gelosa, L. R. Apr., 1973. Water treatment for heating and process steam boilers. *Power Eng.* 77 (4): 54.
(14) Edwards, J. C. and Rozas, E. A. 1961. Boiler scale prevention with EDTA chelating agents. *Proc. Am. Power Conf.* 23: 575.
(15) Edwards, J. C. and Merriman, W. R. 1963. Use of chelating agents for continuous internal treatment of high pressure boilers. *Proc. Int'l. Water Conf., Eng. Soc. W. Pa.* 24: 35.
(16) McCoy, J. W. 1981. *The chemical treatment of boiler water.* New York: Chemical Publishing Co.
(17) Wackenhuth, E. C. and Richards, H. E. 1966. Recent experiences in chemically cleaning a modern boiler. *Proc. Am. Power Conf.* 28: 785.

(18) Rice, J. K. July, 1961. Repeated acid cleaning of boilers. *Combustion* 33 (1): 45.
(19) Engle, J. P. and Keeney, B. R. May 26, 1970. *Prevention of ferric ion corrosion using acid cleaning solution containing hydrogen sulfide and an aldehyde corrosion inhibitor.* U.S. Patent No. 3,514,410.
(20) Figueroa, W. G., Raskowski, R., and Weiss, W. May 24, 1973. Lung cancer in chloromethyl ether workers. *New Eng. J. of Medicine* 288 (21): 1096.
(21) Bloom, D. M. and Gess, L. R. 1980. A new oxygen scavenger for boiler applications. *Proc. Int'l. Water Conf., Eng. Soc. W. Pa.* 41: 269.
(22) Windholz, M., ed. 1976. *The Merck Index.* Rahway, N.J.: Merck & Co.
(23) Adams, R. D. and Chodosh, D. F. 1978. Positively charged ligands. The coordination behavior of sulfonium cations. *J. Am. Chem. Soc.* 100: 812.
(24) Frenier, W. W. and Settinieri, W. J. Oct. 9, 1973. *Sulfonium compounds as corrosion inhibitors in aqueous acid.* U.S. Patent No. 3,764,543.
(25) Frenier, W. W. and Settinieri, W. J. July, 1980. A mechanistic study of sulfonium salts acting as ferric ion corrosion inhibitors. *Corrosion* 36 (7): 323.
(26) Stander, R. A. 1958. Precautionary procedures in chemical cleaning. NACE Publication 8A159.
(27) Noll, D. E. 1964. Factors that determine treatment for high-pressure boilers. *Proc. Am. Power Conf.* 26: 753.
(28) Neat, F. U. March, 1961. The relation between boiler cleanliness and feedwater. *Combustion* 32: 35.
(29) Long, N. A. 1966. Recent operating experience with stainless steel condenser tubes. *Proc. Am. Power Conf.* 28: 798.
(30) Martin, R. C. and Abel, W. T. Nov. 8, 1960. *Copper and iron containing scale removal from ferrous metal.* U.S. Patent No. 2,959,555.
(31) Blume, W. J. Mar., 1977. Citric acid based chemical cleaning processes. *Materials Perf.* 16: 15.
(32) Howell, F. W. and McConomy, T. A. 1966. Maintaining turbine capability through boiler water purification. *Proc. Am. Power Conf.* 28: 808.
(33) Straub, F. C. and Grabowski, H. A. 1945. Silica deposition in steam turbines. *Trans. ASME* 67: 309.
(34) Pocock, F. J. and Stewart, J. F. 1963. The solubility of cop-

per and its oxides in supercritical steam. *J. Eng. Power* 85-A: 33.

(35) Morris, E. B. and Call, R. G. Chemical cleaning of turbines for removal of copper oxide deposits. ASME Paper 64-WA/BFS-1.

(36) Call, R. G. Oct. 10, 1961. *Cleaning composition and a method for its use.* U.S. Patent No. 3,003,970.

VIII

Water-Cooled Heat Exchangers

The term heat exchanger, as used in this chapter, refers to any kind of unfired vessel consisting of tubes within a shell wherein heat is transferred from one fluid to another. Examples are feed water preheaters, surface condensers, waste heat boilers, and process coolers. It is highly desirable that heat exchangers be designed with cooling water in the tubes rather than in the shell, although the latter are by no means uncommon. In order to induce turbulent flow in the shell, baffles are required, but unfortunately these interfere with corrosion inhibition by chemicals in the water, the effectiveness of which depends upon the continual renewal of a passive or protective film on the surface of the metal. Unprotected areas of steel baffles in shells are rapidly corroded by aerated water and are often found to be completely disintegrated after only a few weeks in service. In the process, voluminous deposits of iron oxides are formed that interfere with water circulation and heat transfer. Another serious disadvantage is the low flow rate of water through a shell. This permits suspended solids in the cooling water to settle against baffles where they intensify corrosion by establishing oxygen concentration cells; areas of metal deficient in oxygen are anodic to those exposed to high oxygen tension.

The types of fouling to be expected on the water and process sides of water-cooled heat exchangers are discussed in Chapter I, pp. 1-33. Water-formed deposits include iron tuberculation, insoluble salts of calcium, magnesium, iron, sulfate, and silicate, in addition to hydrous oxides, residues from water treating chemicals, and extraneous materials such as leaves, twigs, fibers, and microbiological growths. Then too, on dry, windy days, especially when earth moving and site preparation are in progress, dirt is carried by the wind into cooling towers and is then pumped with the cooling water into heat exchangers where the velocity is low. Most cooling towers are constructed of wood, parts of which weather and gradually decompose, requiring replacement occasionally. Wood chips and sawdust from these repairs also find their way into the cooling system, lodging sometimes in heat exchanger tubes where they rapidly build up dams that restrict the flow of water. Process leakage in petroleum refining plants frequently introduces hydrocarbons into recirculating cooling systems; in many instances these can be removed by adding a few parts per million of the emulsifying agent sodium dioctylsulfosuccinate to the recirculating water.[1] More serious are leaks of hydrogen sulfide, sulfur dioxide, or alkanolamines, all of which reduce chromate in corrosion inhibitors to the gelatinous hydrous chromic oxide, $Cr(OH)_3$, which has a tendency to block tubes.

Fouling on the process side in petroleum refining, notably in the first stages of hydrocracking and catalytic reforming, often contains large quantities of troilite, FeS, pyrite, FeS_2, or pyrrhotite, $Fe_{(x-1)}S_x$, most often Fe_6S_7 or $Fe_{11}S_{12}$; the latter two differ from troilite and pyrite in being magnetic. Of these sulfides, only FeS is soluble in hydrogen ion, but account must always be taken of its presence because of the rapid evolution of hydrogen sulfide upon adding acids to ferrous sulfide.

$$FeS + 2H^+ = Fe^{++} + H_2S \qquad \text{(VIII-1)}$$

As hazardous heavy metals are seldom encountered in petroleum in any significant amount, the disposal of waste cleaning solution is simplified. Carbon, heavy oily material, and tars, however, are usually present to some degree in process side deposits, and often complicate chemical cleaning. In general, the higher the temperature of a process, the more organic material in the fouling; if more than 30 percent of total carbon is present, chemical cleaning is unlikely to be successful.

VIII.1 COMPOSITION AND DESIGN OF HEAT EXCHANGERS

The choice of a suitable acid, alkali, or other solvent for chemical cleaning is based upon the compatibility of that solution with the materials of construction of the unit to be cleaned, the efficacy of the solvent in dissolving the material with which the unit is fouled, and the cost of the cleaning agent to be used. Table II-8, p. 62, contains a list of solvents, together with a number of metals with which the solvents are either compatible or incompatible. If an incompatible solvent is used—for instance, hydrochloric acid for an austenitic stainless steel—disastrous damage can be done very quickly. For this reason, it is imperative that the materials of construction be positively identified in any unit to be chemically cleaned.

From any group of otherwise acceptable solvents, the most efficient can be selected only by comparing the results of solubility tests on a sample of the fouling to be removed. As a practical matter, however, the principal advantage of chemically cleaning a heat exchanger is that it can be done with the tube bundle in place and without opening the unit. Therefore, a sample is not normally available. This being the case, it is necessary to rely on experience, even though it is difficult to predict results with any accuracy.

Several special precautions are necessary to prevent serious damage to certain alloys during the various steps in chemical cleaning. Copper-based alloys are widely used in heat transfer equipment because of their high resistance to corrosion in both fresh and salt water. As they are susceptible to stress corrosion cracking when exposed to moderate concentrations of ammonia, this chemical must not be used for neutralizing acids in equipment containing admiralty brass, aluminum brass, aluminum bronze, or other copper alloys. Nickel-based alloys including monel, duranickel, and inconel are used for columns, tankage, piping, and heat exchanger tubing in hydrofluoric acid alkylation plants. If not properly stress-relieved, they are subject to stress corrosion cracking in the presence of HF, particularly when it is in the vapor phase. Also, monel is not at all resistant to sour water containing hydrogen sulfide, ammonia, and mercaptans. With a few exceptions, austenitic stainless steels are highly susceptible to stress corrosion cracking when exposed to small concentrations of chloride ion or sulfurous acid. In the usual shutdown procedure in a catalytic reformer plant it is customary to flush the equipment with water, followed by steaming to strip light hydrocarbons, then to apply a final water wash. If, at this point, the warm, moist equipment is opened to the atmosphere, deposits of ferrous sulfide are oxidized to sulfurous acid, which causes severe cracking in stressed stainless steel.

$$2FeS + 7/2O_2 + 2H_2O = Fe_2O_3 + 2H_2SO_3 \qquad \text{(VIII-2)}$$

To preclude this eventuality, blinding, piping, and chemical cleaning, usually with inhibited dilute phosphoric or citric acid, are carried out under nitrogen purge. The acid is succeeded by a rinse with 5 percent sodium carbonate, which is then drained without flushing, leaving a layer of Na_2CO_3 on the stainless steel surfaces to neutralize any sulfurous acid produced by the air-oxidation of sulfides incompletely removed in the acid cleaning. Specific recommendations for shutting down and chemically cleaning catalytic reformer plants have been published by the *National Association of Corrosion Engineers.*[(2)]

In Section VIII.4, below, a general procedure is given that can be applied to nearly any tube-in-shell heat transfer equipment. In industrial steam plants, for example, condensate is often used to preheat boiler feed water in an exchanger perhaps 15 ft long containing a horizontal bundle of 350-400 carbon steel U-tubes. In this type of preheater, hot condensate enters at the top center of the shell, striking an impingement plate provided to minimize erosion on the upper surfaces of the tubes, then flows horizontally toward the two ends, while being confined to the upper half of the tube bundle by a horizontal pass partition. When the condensate water reaches the ends of the shell, matters are so arranged that it then returns through the bottom half of the bundle to the center of the shell, where it exits from the bottom. There is seldom any need to clean the shell side of this type of preheater, although the outsides of the tubes in the center section of the bottom half of the bundle often show rather severe carbon dioxide grooving, as well as some pitting. This is the place where the condensate is the coolest, so consequently the solubility of carbon dioxide is highest and the pH is lowest. The insides of the tubes may become fouled with iron oxides if the feed water is improperly deaerated.

A surface condenser is essentially a closed shell filled with many tubes through which cooling water circulates. Exhaust steam from a turbine is piped into the shell where it condenses on the tubes containing cold cooling water. In this process a large volume of steam is condensed to a small volume of water, producing a vacuum in the shell and reducing the backpressure on the turbine, thereby increasing its efficiency. The vacuum in the shell is stabilized and noncondensible gases are removed by passing additional steam—typically, at 150 psi—through eductors attached to the shell. If the eductor nozzles or the tubes containing cooling water become fouled, additional steam must be used to obtain the required vacuum. Because of the very large volume of condensers in utility plants, it is advisable to clean their shells by hand. For the same reason, surface condenser tubes are most efficiently cleaned by foamed acid, prepared by blowing nitrogen into an acid solution containing a surfactant. Nitrogen gas accounts for 70–80 percent by volume of the resulting foam, which completely fills bent tubes, and transports scale and fines, often more effectively than does an acidic solution. Further advantages of foaming are a saving in time of about one-third over liquid cleaning and a much smaller volume of liquid is left to be disposed of after the cleaning. Acid foaming is discussed further in Chapter III, pp. 92–93; it is inapplicable in the presence of hydrocarbons, for they destroy the foam.

The majority of coolers used in petroleum refining and chemical processing are shell-and-tube-type exchangers. In fixed-tube-sheet heat exchangers the tubes are rolled into the sheets, one at each end of the exchanger, and both are welded to the shell. In an internal-floating-head exchanger, only one tube sheet is welded to the shell and the other, closed by a floating-head cover, floats freely in the shell to accomodate differential thermal expansion.(3) The same end can be accomplished by using U-tubes with both ends rolled into the same tube sheet, the disadvantage being that U-tubes are more difficult to chemically clean than are straight tubes. A special procedure, described in Section VIII.5, below, is necessary when water-jetting floating-head heat exchangers.

VIII.2 CONDITION OF THE EQUIPMENT

Most instances of fouling in water-cooled heat exchangers can be attributed to two causes: (1) failure to back-flush exchangers regularly while in service; (2) failure to passivate new or chemically cleaned steel exchangers properly. If either of these two points is neglected, voluminous deposits of iron oxides inevitably form and heat transfer coefficients decline. Eventually, tubes may become completely plugged, but this can only be ascertained by removing the channel covers and inspecting the tubes. It is important to do this, if at all possible, as chemical cleaning cannot be successful unless the solution can circulate. Furthermore, the failure of a chemical cleaning will not be apparent until the unit is put back in service, with little or no improvement in performance. Opening an exchanger for inspection also makes possible the collection of a sample of the fouling and the detection of any oil or other organic material that must be removed before attempting an acid cleaning.

When a sample of fouling is available, qualitative solubility tests can readily be made by introducing a small portion of the sample into approximately 100 ml of each of several possible solvents, then heating at use temperature for 4–6 h. After the heating period, visual inspection of the solutions indicates what can be expected in actual use. Complete dissolution is seldom achieved or necessary. Most often binders that hold fouling material together are dissolved, and the remaining loosened insoluble material can be flushed out of the exchanger readily. Occasionally, successive solvent treatments are necessary; when organic matter is present an alkaline wash must always precede acid treatment.

The bulk of a water-formed deposit is, in most instances, mixed oxides of iron or, in systems where pH control is erratic, calcium carbonate. Both of these can be removed from carbon steel, brass, or nickel alloy heat exchangers by circulating a suitably inhibited dilute solution of hydrochloric acid; dilute phosphoric acid or one of the organic acids described in Chapter II should be used in stainless steel equipment. In Section VIII.4, below, a general procedure is given for cleaning a carbon steel heat exchanger shell with hydrochloric acid.

Whenever bulky deposits of iron oxides are encountered, it will be found that only part of the accumulation dissolves and that a considerable mass of oxides remain when the unit is drained after an acid cleaning, even though acid spending has ceased and the concentration of dissolved iron in the recirculating solution has reached a constant value. The reason for this is that iron oxides are not rapidly dissolved by dilute hydrogen ion unless they are in contact with iron. That is, the reaction represented by Eq. (VIII-3) is very slow in dilute acid solutions, whereas that in Eq. (VIII-4) is rapid.

$$Fe_2O_3 + 6H^+ = 2Fe^{+3} + 3H_2O \qquad \text{(VIII-3)}$$

$$Fe_2O_3 + Fe + 6H^+ = 3Fe^{++} + 3H_2O \qquad \text{(VIII-4)}$$

Thus, at the beginning of a chemical cleaning there is a relatively large area of rusty steel from which ferric oxide and some of the steel itself immediately dissolve. As the procedure continues, however, undissolved oxides tend to accumulate into heaps, as is often observed in water wall headers in boilers and in channels and piping in heat exchangers. As the reaction shown in Eq. (VIII-3) is slow, these piles of rubble are not much affected by the acid, even though the accumulation is saturated with it. When a very large quantity of rust is encountered, as in tube-in-shell waste heat boilers, for example, it is often difficult to neutralize all of the acid in these vessels, especially when cleaning the shell. It is more-or-less analogous to removing an acid solution from a sponge by washing its surface with water. Regardless of the difficulty, however, the cleaning contractor must be absolutely certain that all of the acid is neutralized before adding a sodium nitrite passivating solution, otherwise the very toxic gas, NO_2, is immediately released. Neutralizing and rinsing can be accomplished much more quickly if a wetting agent is used, or if a 3/8-in. fitting is provided on the bottom of the shell so that the solution and insoluble matter can be agitated by sparging nitrogen into it at, say, 175-250 scf/min. Nitrogen should not be introduced during the acidification step or the loss of metal will be excessive.

In view of the foregoing interpretation, the question of ferric ion concentration is moot, because the reaction in Eq. (VIII-4) occurs almost instantaneously whenever any appreciable concentration of ferric ion is generated from any source. It is likely that many ferric ion values reported in cleaning solutions from steel vessels are erroneous results arising from deficiencies in the analytical methods used. A few other comments on this subject are made in Chapter III , pp. 74-75.

If a heat exchanger contains a substantial amount of loose rubble in the tubes or channels, the shell covers (often called heads) and the channel covers can be removed, and the insoluble material ejected by high-pressure water jetting. As a rule, tubes can also be more effectively cleaned this way than by circulating an acid. If the tube bundle is fouled on the shell side, however, it must be removed from the shell in order to be water jetted. This may necessitate preliminary chemical treatment, such as filling the shell with hot, alkaline emulsion and allowing it to soak for several hours before the tube bundle can be freed from the shell. In any event, a crane and a rigging crew are required to pull the bundle out, which adds a great deal to the expense of the cleaning operation, although this may be offset by the savings in cost of disposing of a large volume of hazardous waste. Incidentally, it must always be borne in mind that triangularly pitched tubes cannot be cleaned by water jetting. More details of the procedure are included in Section VIII.5.

VIII.3 GENERATION OF TOXIC GASES

The possible presence of deposits of ferrous sulfide is always a consideration when contemplating the acid cleaning of the process side of water-cooled heat exchangers in the first stage sections of hydrocracking and catalytic reforming plants. In these two processes, which are carried out in the presence of excess hydrogen, all sulfur compounds in the feed stock are reduced to H_2S, while nitrogen compounds are reduced to NH_3. In catalytic cracking processes, sulfur compounds are also reduced to H_2S, but nitrogen compounds are frequently transformed to HCN. Because of the corrosive effect of H_2S on steel, FeS is nearly always found in corrosion deposits in heat exchangers in plants where these catalytic refining processes are carried out.

$$Fe + H_2S = FeS + H_2 \qquad \text{(VIII-5)}$$

When an acid is brought into contact with FeS, H_2S is immediately evolved, as shown in Eq. (VIII-1). As this gas is highly toxic, and forms explosive mixtures with air, provision must be made for protecting workmen and for containing the gas so that it can be disposed of safely. Deposits containing cyanides release HCN when contacted by acids. Hydrogen sulfide at high concentrations (1000–2000 ppm) causes immediate death by asphyxiation, resulting from respiratory paralysis; hydrogen cyanide, an enzyme poison, also causes death by asphyxiation.

Because of the rapid onset of olfactory fatigue, the odor of H_2S is an unreliable indicator of its concentration. Upon being inhaled at concentrations above 70 ppm, irritation of the lungs and mucosa occurs, whereas exposure to 250–600 ppm causes such symptoms as headache, dizziness, nausea, gastrointestinal disturbances, dryness and pain in the nose, throat, and chest, and coughing. Individuals repeatedly exposed to H_2S become sensitized, causing symptoms to become more severe; chronic poisoning is not unusual.[(4)] The chemical cleaning contractor must be thoroughly familiar with OSHA regulations governing the handling of hydrogen sulfide and the safety equipment required, before undertaking any cleaning in which the gas may be released.

Cyanides are rapidly absorbed by inhalation, ingestion, or through the skin. Exposure to hydrogen cyanide at concentrations greater than 10 ppm results in irritation of the lungs, a burning sensation on the tongue, a metallic taste, weakness, headache, dizziness, nausea, vomiting, difficulty in breathing, heart arrythmias, convulsions, coma, and death. Sudden exposures to high concentrations may cause immediate collapse, convulsions, and death.[(4)]

Tewari and Campbell[(5)] have studied the rate of evolution of H_2S from FeS at various values of pH, and found that the rate of dissolution is determined by the surface reaction between hydrogen ion and FeS. Thus, the rate is increased only by increasing the concentration of H^+; for this reason, maintaining a high rate of circulation has much less effect on the

$FeS\text{-}H^+$ reaction [Eq. (VIII-1)] than it does in the dissolution of most other deposits. The rate law for the reaction is,

$$\frac{d(H_2S)}{dt} = K_f[H^+] - K_r[Fe^{++}]^{1/2}[P_{H_2S}]^{1/2} \qquad \text{(VIII-6)}$$

Lawson and Martin,[6] in investigating several different procedures for dissolving ferrous sulfide, confirmed that the value of K_f, the rate constant for the dissolution reaction in acid is $K_f = 2.7 \times 10^{-3} \mathrm{l m^{-2} s^{-1}}$. The reverse reaction rate constant, K_r, is not significant when $[H^+]$ is large, as it is in a chemical cleaning solution.

As both H_2S and HCN are acidic gases, they can be absorbed in a single solution of sodium hydroxide, provided care is taken to maintain free alkali in the solution at all times. The chemical equilibria in alkaline sulfide solutions are more complicated than assumed here, but as a first approximation the reactions can be formalized in steps. The reader interested in a more exact treatment and precise analytical procedures, is referred to an earlier publication.[7]

When acid is added to ferrous sulfide, the hydrogen sulfide is transferred to a scrubber containing a solution of sodium hydroxide. At the outset, when sodium hydroxide is in large excess, H_2S is neutralized to S^{--}. As the concentration of OH^- decreases, HS^- begins to form, and if the concentration of OH^- were allowed to reach zero, H_2S would then pass unchanged through the scrubber. The steps are,

$$FeS + 2H^+ = Fe^{++} + H_2S \qquad \text{(VIII-1)}$$

$$H_2S + 2OH^- = S^{--} + 2H_2O \qquad \text{(VIII-7)}$$

$$H_2S + OH^- = HS^- + H_2O \qquad \text{(VIII-8)}$$

In addition to the above, the following equation can be written:

$$H_2S + S^{--} = 2HS^- \qquad \text{(VIII-9)}$$

which points up one of several complications in the chemistry of alkaline sulfide solutions, namely, the high degree of hydrolysis of S^{--} owing to the very small value of the second dissociation constant of H_2S.

$$S^{--} + H_2O = HS^- + OH^- \qquad \text{(VIII-10)}$$

The hydrolysis constant of this reaction is,

$$K_h = \frac{(HS^-)(OH^-)}{(S^{--})} = \frac{K_w}{K_2} = 10^{-14}/7.1 \times 10^{-15} = 1.4 \qquad \text{(VIII-11)}$$

Eq. (VIII-11) shows that HS^-, OH^-, and S^{--} can, and do, coexist in alkaline solutions. Table VIII-1 summarizes the ions in solution in several different pH ranges.

TABLE VIII-1

Ionic Species in Alkaline Scrubbers

pH range	*Ions in solution*
13–14	HS^-, S^{--}, OH^-
12–13	HS^-, OH^-
9–12	HS^-
5–9	H_2S, HS^-

Table VIII-1 suggests the possibility of controlling the concentration of free alkali in scrubbers by measuring pH. The magnitude of the sodium ion error, however, makes the precise determination of pH very difficult in highly alkaline solutions.(8) Titration procedures are given in Chapter XI that are adequate for estimating free alkali in such solutions.

At this point it might be well to explain the much-misused term, caustic availability. This value, which unfortunately is expressed in percent, means the percentage of the total alkalinity in a solution that has the capacity to neutralize H_2S, or some other acid. It is determined by titrating a sample to phenolphthalein (free alkalinity), then on to the methyl orange end point (total alkalinity). The free alkalinity titration divided by the total alkalinity titration, then multiplied by 100, is percent available caustic. Note that the value obtained is independent of sample size, so it tells nothing about the actual neutralizing capacity of a solution; it merely means that however much alkalinity is present, a certain percentage of it is free. For example, although a 0.1 N solution and a 50 percent solution of NaOH both have 100 percent availability, a gallon of the second solution contains 190 times as much free OH^- as a gallon of the first solution.

Neither the term nor the procedure should be used in controlling the concentration of a caustic scrubber; percent free NaOH is what is required.

As the weight of FeS in an exchanger can never be known in advance, it is necessary somehow to ensure that the caustic scrubber does not become exhausted. This can be done by balancing the weight of NaOH in the scrubber with the weight of acid added in the cleaning solution, so that the percentage of free NaOH in the scrubber solution does not fall below 10 percent. The method for doing this is best illustrated by a numerical example. Densities, percentages, and concentrations quoted in the following development and in Section 4.b, below, can be obtained from a chemical handbook.(9)

If a 1000-gal scrubber tank is available, a convenient volume of scrubbing solutions is 600 gal. A single scrubber is sufficient, if its vent can be piped to a low-pressure relief system, but, if not, a secondary scrubber, vented to the atmosphere, is necessary. It is much easier to purchase concentrated solutions of caustic soda than to attempt to prepare them from flakes. The concentrated solution is diluted in the scrubber to prevent the precipitation of Na_2S in the jets of the sparger, and to provide a larger volume of solution for the gas to pass through so that absorption is more likely to be complete. Vents should be protected by demister pads.

Suppose that the objective is to clean the shell of a heat exchanger that is estimated (or guessed) might contain 1000 lb of FeS. The scrubber is charged with 250 gal of 50° Bé sodium hydroxide, which is then diluted to 600 gal with water. As 50° Bé caustic contains 6.365 lb of NaOH per gallon, the scrubber contains $250 \times 6.365 = 1591$ lb of NaOH. The diluted solution contains $1591/600 = 2.65$ lb of NaOH per gallon, with a gravity of ~ 31° Bé and a strength of ~ 25 percent. As a safety precaution, the caustic should not be neutralized to less than 10 percent of free NaOH by H_2S. A 10 percent solution contains 0.925 lb per gal of NaOH, so 600 gal contains $0.925 \times 600 = 555$ lb of NaOH. The weight that is permissible to neutralize is, thus, $(1591 - 555) = 1036$ lb of NaOH.

In the actual cleaning operation, HCl is replaced by H_2S, which then neutralizes the NaOH, but as the amount of H_2S released is equivalent to the acid added [cf. Eqs. (VIII-1) and (VIII-7)], the amount of acid consumed can be calculated from Eq. (VIII-12).

$$NaOH + HCl = NaCl + H_2O \qquad \text{(VIII-12)}$$

This equation shows that one mole of NaOH neutralizes one mole of HCl, or 40 lb of NaOH neutralizes 36.5 lb of HCl. Therefore, $36.5 \times 1036/40$

= 945 lb of HCl neutralizes 1036 lb of NaOH. The volume of various strengths of concentrated hydrochloric acid solutions required can be found in Table VIII-2, in which the last column shows the number of gallons of each strength of HCl that will neutralize 600 gal of 25 percent NaOH to 10 percent free caustic.

TABLE VIII-2

Strength of HCl Solutions

%	lb HCl/gal	gal/945 lb HCl
10	0.874	1081
28	2.66	355
32	3.10	305
37	3.66	258

Again, hydrochloric acid neutralizes sodium hydroxide indirectly by liberating hydrogen sulfide. Incidentally, 945 lb of HCl can dissolve approximately 1140 lb of FeS, liberating in the process 440 lb of H_2S, equivalent to some 4650 standard cubic feet of the toxic gas. Obviously, if there is less FeS than anticipated, less NaOH and HCl will be consumed.

Sodium arsenite has been used to inhibit corrosion in gas conditioning plants where H_2S and CO_2 are removed by absorption in monoethanolamine or diethanolamine.[10] Uhlig[11] suggests that corrosion protection is achieved by the reduction of arsenite to arsenic at cathodes; the element has a high overvoltage. Various compounds, including Fe_3O_4, $FeCO_3$, β-Cu_3As, NiAs, Cu_2S, and Cu_3AsS_4 have been found in fouling from monoethanolamine coolers.* Upon acidification, some of these evolve the very hazardous gas arsine, AsH_3. Sax[12] reports that the principal physiological effect of arsine is to hemolyze red blood cells with ensuing jaundice, anemia, and renal damage; it is also a known carcinogen. Elkins and Fahy[13] have reported two cases of arsine poisonong resulting from the cleaning of aluminum tanks.

Should occasion arise to clean any part of an alkanolamine system that has ever been inhibited with arsenite, it is essential to provide for the cap-

* J. Q. Adams, Chevron Research Company, private communication.

ture of arsine. Caustic scrubbers, which must be used to absorb H_2S, do not retain AsH_3, so it is necessary to add an after-scrubber containing permanganate solution. As arsine is a very powerful reducing agent, it is readily oxidized by any moderate oxidizing agent.

$$8MnO_4^- + 3AsH_3 + OH^- = 8MnO_2 + 3AsO_4^{-3} + 5H_2O \qquad \text{(VIII-13)}$$

A consideration of molecular weights in Eq. (VIII-13) shows that 2.03 lb of $KMnO_4$ are required to oxidize 1 lb of AsH_3. In most instances, 100–200 lb of $KMnO_4$ in a 3–5 percent solution is sufficient; the purple color of the absorbing solution is observed freqently during the cleaning operation to make certain that it does not become exhausted.

VIII.4 TYPICAL CHEMICAL CLEANING PROCEDURE

At this point a specific procedure can be developed for cleaning a tube-in-shell heat exchanger. In the first stage of the catalytic reforming process, heavy gasoline is hydrogenated to saturate any olefins and unsaturated naphthenes, and to convert thiophenes, mercaptans, and other sulfur-containing compounds to hydrogen sulfide. The first-stage reaction product passes through a feed-effluent exchanger where it is cooled by the heavy gasoline feed. Quench water is next injected to extract NH_3-H_2S, and the mixture is then further cooled by being passed through the shell of an internal-floating-head heat exchanger with cooling water in the tubes. In the course of time, iron sulfide deposits accumulate in the shell because of the corrosive action of the sour water mixed with the first stage product. In addition, scale, corrosion products, dirt, and other debris can block the water tubes. It is proposed that the shell be chemically cleaned with the tube bundle in place, and that the tubes be cleared by high-pressure water jetting, as described in Section VIII.5.

The shell of the unit to be cleaned is 23 ft long, has an inside diameter of 4 ft, and has a capacity of 1400 gal with the tube bundle in place. The capacity of the tubes and channels is 800 gal. The shell and tubes are carbon steel, so they can be cleaned with inhibited hydrochloric acid. Before any acid can be effectively used, however, the shell must be deoiled. As the maximum boiling point of catalytic reformer feed stock is limited to 400°F (to exclude polynuclear aromatics), heavy organic material is seldom encountered at this point, and deoiling can usually be accomplished by circulating a solution of trisodium phosphate containing a small amount of a nonionic wetting agent through the shell. Following this, inhibited hydrochloric acid is circulated through the unit and the evolved H_2S is absorbed in sodium hydroxide, transferred to a low-pressure relief system, or piped to a flare. The latter procedure is environmentally objectionable, as H_2S is oxidized to sulfur dioxide, SO_2, in the flaring. Finally, the shell is rinsed, neutralized, and passivated.

VIII.4.a Deoiling

When shutting down hydrocarbon processing units for inspection or cleaning, it is customary, as a safety precaution, first to wash them with water, next to steam them to volatilize light hydrocarbons, and then to water wash again. If a unit is not to be opened, and steaming can be omitted, lighter hydrocarbons remain that increase the solubility of heavy gas oil contamination. Polymeric material cannot be emulsified by the procedure suggested here; an alkaline emulsion, as described on p. 40, should be applied instead. Also, certain antifoulants added to crude petroleum may increase the percentage of organic material in some heat exchangers as the result of the formation of degraded polymers. These, too, yield to alkaline emulsions.

Before releasing a unit to the chemical cleaning contractor, the entire system, including blinds and temporary piping, should be filled with water and hydrostatically tested by the client. If the unit is turned over to the contractor filled with water, his leak testing can be completed in just a few minutes. The schematic diagram in Fig. VIII.1 shows the the essential cleaning components and their arrangement. Although this drawing gives the impression of representing a rather extensive system, it is in reality very compact. The circulation pump, acid tank, supply tank, and surge tank are all mounted on a single truck. The caustic scrubber may be supplied by either the client or the contractor; in the latter case one, or more often two, 1000-gal tanks are mounted on a second truck positioned close to the chemical cleaning truck to permit a short-connection with the surge tank. The trucks are both located as closely as possible to the unit being cleaned so that pipe or hose runs are short.

Before the contractor can connect his equipment to the unit to be cleaned, he must be provided with a detailed piping diagram and blind list, which he must verify himself. Fig. VIII.2 shows the piping arrangement for cleaning the shells of either or both of two heat exchangers, A and B. As contractors customarily use 2-in. pipe or hoses for connecting their equipment, it is necessary for the client to provide 2-in. fittings on units to

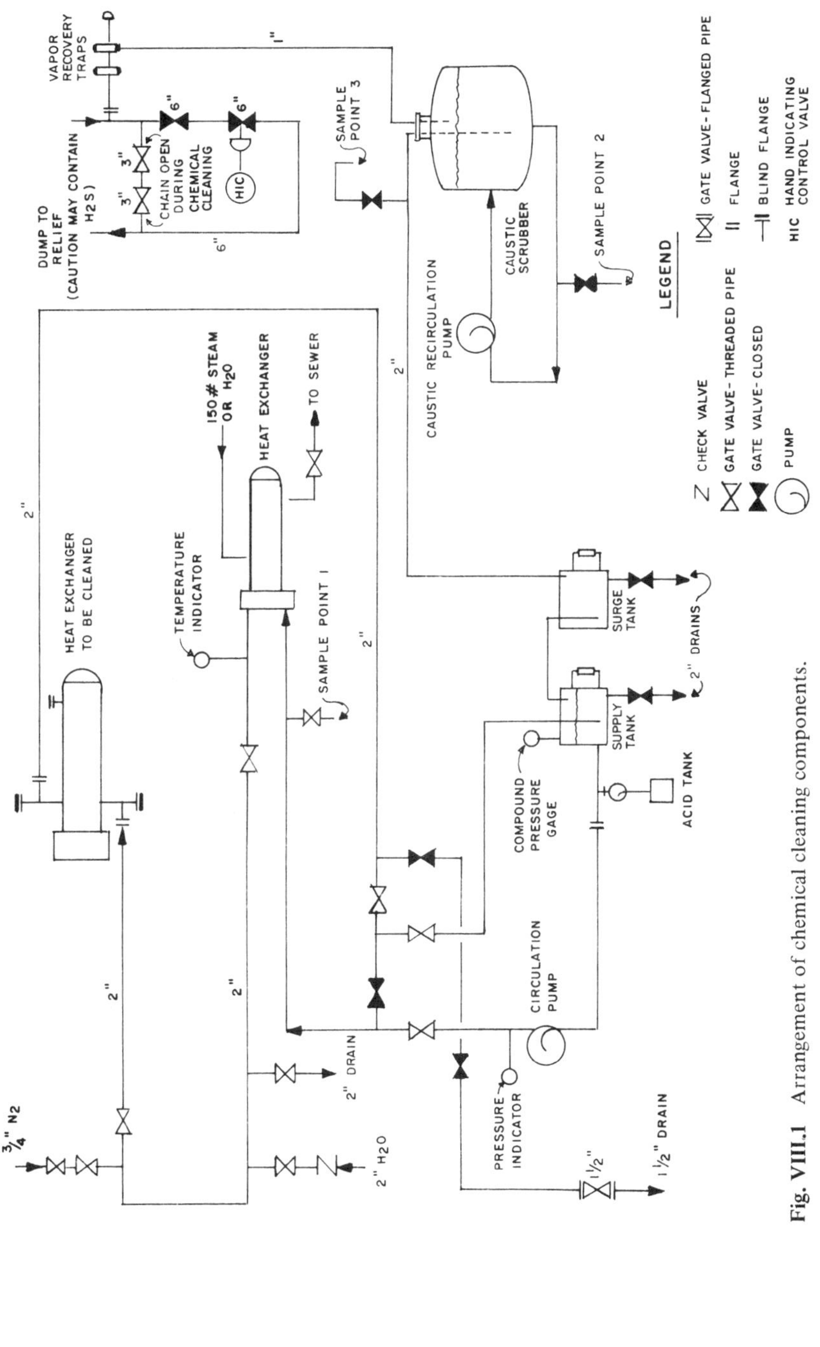

Fig. VIII.1 Arrangement of chemical cleaning components.

be cleaned. In order to get sufficient cleaning solution into exchanger A through existing fittings, it is necessary to install a 2″ × 1″ swedge at position ① and two 2″ × ¾″ swedges at position ② in Fig. VIII.2. Thus, three 2-in. connections are needed between the circulating pump manifold and the heat exchanger inlets. To avoid this inconvenience, a single connection was subsequently installed in the 6″ line leading to the lower shell nozzle of exchanger A. A section of pipe, usually a spool, is removed at position ③ on both A and B, and replaced with a 2″ valve and nipple to connect to the return line to the supply tank on the chemical cleaning truck. Also, a 2″ valve and nipple are installed at ④ to provide an inlet to exchanger B. The contractor must make sure that all blinds are in place and that pipefitting has been completed as shown in the diagram supplied by the client, before making connections.

The following deoiling procedure is, in most instances, effective. Assuming the system is full of water, drain about 300 gal from the supply tank to the sewer, then add sufficient chemicals to produce a solution containing 2 percent of trisodium phosphate, 0.25 percent of sodium hydroxide, and 0.1 percent of a nonionic wetting agent. The sodium hydroxide is not really necessary, but it is customarily included when sulfides may be present.

As the shell has a capacity of 1400 gal of water and the 1000-gal supply tank contains 700 gal of water, there is a total of 2100 gal plus a few more in the piping. This is approximately 17,500 lb of water, from which it can readily be calculated that 17.5 lb of wetting agent, 44 lb of NaOH (~ 7 gal of 50° Bé caustic soda), and 350 lb of Na_3PO_4 (or 815 lb of $Na_3PO_4 \cdot 12H_2O$) are required. The trisodium phosphate is added to the supply tank and dissolved by agitating with steam, after which the sodium hydroxide and wetting agent are added.

The alkaline solution is heated to 180°F and circulated at about 750 gpm for 3 h. In the arrangement shown in Fig. VIII.1, the solution in the supply tank is pumped to a heat exchanger where it is heated to the desired temperature by 150-psi steam. The warm solution passes into the heat exchanger to be deoiled through the bottom shell nozzle, flows out of the top shell nozzle, and returns to the supply tank. Although no H_2S is evolved at this stage, the scrubber(s) should be connected and in place from the beginning.

At the end of 3 h the flow of solution is reversed, holding the temperature at 180°F, and circulating for an additional 2 h. The discharge of the circulating pump is equipped with a reversing manifold to reverse the

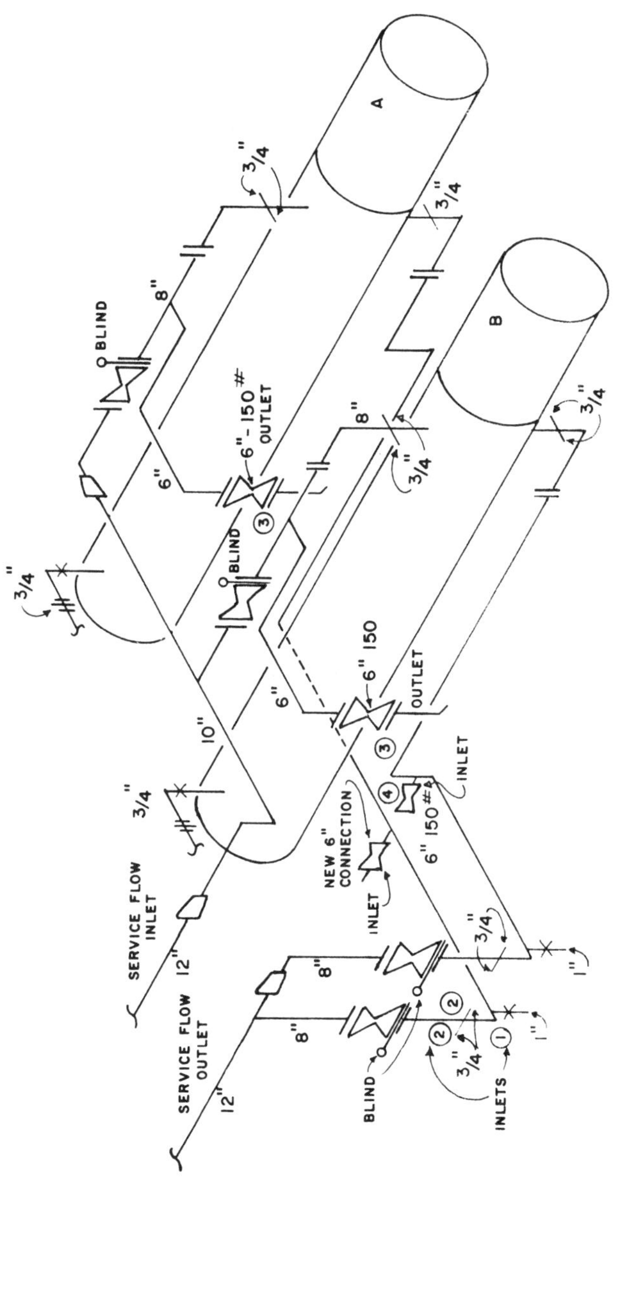

Fig. VIII.2 Detail of cleaning connections to exchangers. ① Install a 2" X 1" swedge for an inlet connection; ② Install two ¾" valves and two 2" X ¾" swedges for inlet connections; ③ Remove 6" pipe and install 2" valve and nipple for outlet connection; ④ Remove 6" boilermaker blind and install 2" valve and nipple.

direction of flow and to allow flushing or draining to a vacuum truck to be accomplished without disconnecting the supply or return lines.

The alkaline phosphate solution is next pumped out of the vented system into a vacuum truck; it can be transferred to an oxidation pond, if one is available, or else trucked to a suitable waste disposal site. The shell is finally flushed with water introduced through a connection on the circulating pump.

VIII.4.b Acid Cleaning

CAUTION: It is most important that the reader be aware that adding acid to sulfide deposits is an extremely hazardous procedure, and that once acid cleaning has been started, and hydrogen sulfide is being evolved, it is impossible to suspend operations safely.

Before admitting any acid to the chemical cleaning loop the amount of it that will exhaust the caustic scrubber to 10 percent free NaOH must be calculated. Two scrubber arrangements are considered: one in which a single scrubber is used with its vent piped to a low-pressure relief system as in Fig. VIII.1; and one in which two scrubbers are connected in series, with the second vented to the atmosphere.

It is assumed that 450 gal of 50° Bé caustic soda, diluted to 600 gal with water, will be sufficient to absorb all of the H_2S liberated from an exchanger of the size being used as an example. As it is necessary to maintain a reserve concentration of 10 percent free NaOH in the scrubber, the total hydrochloric acid eventually admitted must be limited to that amount that leaves the desired concentration of free NaOH in the scrubber solution.

The total weight of NaOH added in 450 gal of a 50° Bé solution is,

$$6.365 \text{ lb NaOH/gal} \times 450 \text{ gal} = 2864 \text{ lb NaOH}$$

The weight of NaOH in 600 gal of a 10 percent solution is,

$$0.925 \text{ lb NaOH/gal} \times 600 \text{ gal} = 555 \text{ lb NaOH}$$

The weight of NaOH that can safely be neutralized by hydrochloric acid is, thus,

$$(2864 - 555) = 2309 \text{ lb NaOH}$$

From the molecular weight relations expressed in Eq. (VIII-11), the weight of HCl that can neutralize 2309 lb of NaOH is,

$$2309(36.5/40) = 2107 \text{ lb HCl}$$

Assuming that a 32 percent solution of HCl is used, the volume required to contain 2107 lb of HCl is,

$$2107/3.10 = 680 \text{ gal of 32 percent HCl}$$

If this amount were continuously blended into water circulating in the system (a total volume of ~ 2400 gal), the concentration of HCl would be,

$$2107 \text{ lb HCl}/2400 \text{ gal} = 0.878 \text{ lb HCl/gal}$$

which is equivalent to slightly more than 10 percent of HCl, assuming no reaction occurred. In practice, sufficient inhibited acid is initially added to produce a solution containing 1-2 percent, i.e., 65-130 gal of 32 percent HCl added to 2400 gal of water circulating through the shell of the heat exchanger and the remainder of the cleaning loop. As pointed out on p. 166, the rate of evolution of H_2S from FeS depends only upon the concentration of the acid; this concentration is kept low initially to prevent too rapid an evolution of H_2S.

Case I. *Single scrubber vented to a relief system.*

Transfer 450 gal of 50° Bé caustic soda to the caustic scrubber, and dilute to 600 gal with cold water. Connect the inlet of the scrubber to the surge tank on the chemical cleaning truck, and the outlet to a low-pressure relief system.

Add approximately 1700 gal of water to the system, and circulate it at 300-500 gpm. Then, *without heating*, and with the valve closed between the supply and the surge tanks, admit about 100 gal of inhibited 32 percent HCl, and observe the pressure in the supply tank. When this pressure reaches 2-3 psi, slowly open the valve to the surge tank. Should the pressure increase significantly (> 5 psi), stop circulating the acid solution, i.e., turn the circulating pump off, until the pressure stabilizes at 2-3 psi.

Using the procedures given in Chapter XI, check for percent HCl at sample point No. 1 in Fig. VIII.1, frequently at first, then every 30 min as cleaning proceeds. When the rate of acid spending decreases, turn on the steam to the heat exchanger, and heat the recirculating solution to 140°F, maintaining this temperature while the remainder of the hydrochloric acid is added.

Stander[14] has pointed out that hydrogen sulfide reduces the efficiency of corrosion inhibitors markedly. Thus, the corrosion rate of carbon steel in inhibited 10 percent HCl is 0.5 mg/cm^2-h (0.025 mph), but if H_2S is added the rate increases to 13 mg/cm^2-h (0.65 mph). For this reason, the temperature of the cleaning solution must not be permitted to exceed 140°F, and the duration of the treatment should be limited to a maximum of 6 h; in general, the process can be completed within 4–6 h. Sulfide stress corrosion, essentially the same phenomenon as hydrogen embrittlement, is also a possibility when H_2S is present, particularly in steels of high tensile strength, e.g., 250,000–270,000 lb/in.^{2} As hydrogen cracking does not occur at temperatures below 400°F, it is not a consideration in chemical cleaning.

Check the percent of free NaOH at sample point No. 2 every 30 min during the cleaning, and discontinue adding acid if the concentration falls below 10 percent. (This cannot happen if the volume of HCl is calculated correctly.) As ferric ions are rapidly reduced by hydrogen sulfide, there is no need to attempt to measure Fe^{+3} when H_2S is present.

$$2Fe^{+3} + H_2S = 2Fe^{++} + S + 2H^{+} \qquad \text{(VIII-14)}$$

It is of some interest to calculate the volume of FeS that can be dissolved by 2107 lb of HCl. The weight dissolved is found from Eq. (VIII-1) to be,

$$2107(88/73) = 2540 \text{ lb FeS}$$

Noting that the density of FeS is 4.84 g/ml, the volume of 2540 lb of it is easily calculated.

$$\frac{2540 \text{ lb} \times 454 \text{ g/lb} \times 0.0353 \text{ ft}^3/\text{l}}{4.84 \text{ g/ml} \times 1000 \text{ ml/l}} = 8.4 \text{ ft}^3$$

It is unlikely that this much FeS will be found in a heat exchanger of the size under consideration.

Case II. *Two scrubbers in series, the second vented to the atmosphere.*

In this arrangement, shown in Fig. VIII.3, the intent is to capture most of the H_2S in the primary scrubber, with the secondary serving as a safety device should the initial evolution of H_2S be too rapid for complete reaction in the primary scrubber. Following is a suitable distribution of sodium hydroxide for the same weight of HCl assumed in Case I, i.e., 2107 lb.

Primary: 375 gal of 50° Bé × 6.365 lb NaOH/gal = 2387 lb NaOH

Secondary: 100 gal of 50° Bé × 6.365 lb NaOH/gal = 636 lb NaOH

The weight of NaOH neutralized by 2107 lb of HCl is,

$$2107(40/36.5) = 2309 \text{ lb NaOH}$$

The weight of unneutralized NaOH in the primary scrubber after all of the HCl has reacted is,

$$(2387 - 2309) = 78 \text{ lb NaOH}$$

If the NaOH solutions in both scrubbers are diluted at the outset to 600 gal, the primary will contain 1-2 percent free NaOH and the secondary will contain approximately 11 percent of free NaOH after all of the HCl has reacted with FeS. In some instances, the caustic in the secondary scrubber can be reused for scrubbing H_2S, after checking it for the total sulfide and free NaOH concentrations.

The procedure for adding acid is carried out as described under Case I. The recirculating acid solution is monitored for percent HCl and percent total iron, while the caustic scrubber(s) are checked for percent free NaOH until such time as these values stabilize, indicating that no further chemical reactions are taking place. The atmospheric vent should also be checked frequently for H_2S using a suitable tester (see p. 00). Should a break-through or leak occur, the area must be cleared immediately while personnel equipped with self-contained breathing apparatus find and repair the leak.

When acidity, free caustic, and total iron concentrations become reasonably constant, admit nitrogen gas at 175-250 scf/min to the system to sweep residual H_2S from the piping and vapor spaces into the scrubber(s). Progress can be assessed by measuring ppm H_2S at sample point No. 3 in Fig. VIII.1, or F in Fig. VIII-3. When the concentration of H_2S here falls below 10 ppm (the OSHA Threshold Limit Value), the contents of the caustic scrubber(s) may be transferred to a vacuum truck. This is done by venting the scrubber to the atmosphere (to avoid collapsing the vessel), then connecting the scrubber to the tank on the vacuum truck. Note that an auxiliary scrubber must be attached to the exhaust from the vacuum pump to absorb any H_2S that comes out of solution when the pressure on it is reduced. As a large volume of air and nitrogen pass through this scrubber, the container must be large enough to prevent splashing or over-

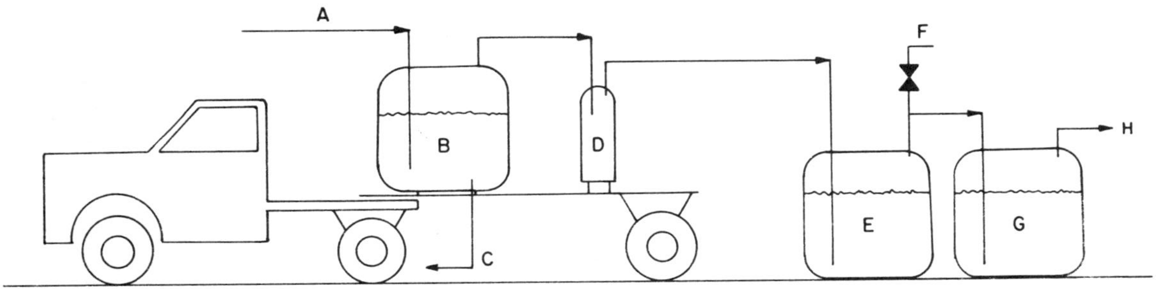

Fig. VIII.3 Chemical cleaning setup with caustic scrubbers. A = Return line; B = Supply tank–1000 gal; C = Supply line; D = Surge tank–350 gal; E = Primary caustic scrubber–1000 gal; F = Sample point for H_2S; G = Secondary scrubber–1000 gal.

flow of the alkaline solution. In petroleum refineries, alkaline sulfide solutions can be disposed of in H_2S recovery systems or in a sulfide oxidizer; in default of these facilities, the solutions can be trucked to an approved waste disposal site.

The spent hydrochloric acid can be neutralized in a vacuum truck containing sufficient caustic soda to accomplish this end. Suppose, for instance, that the 2400 gal of cleaning solution contains 4.0 percent HCl at the conclusion of the acid cleaning. This amounts to $2400 \times 0.34 = 816$ lb of HCl to be neutralized. The amount of NaOH needed to do this is $816(40/36.5) = 894$ lb of NaOH, which is contained in $(894/6.365) = 140$ gal of 50° Bé caustic soda. This volume is introduced into an empty vacuum truck, and the spent HCl is then transferred to the truck. The neutralized solution can be disposed of without difficulty, once its pH has been adjusted to 5-9, preferably by adding soda ash, Na_2CO_3, which has the advantage of exerting a buffering effect.

WARNING: Alkaline sulfide solutions must never be mixed with acidic solutions. Particular attention must be paid to the sequence of steps when transferring these solutions to vacuum trucks. Also, bear in mind that pockets of toxic gas can remain trapped in the system, even after chemical cleaning is finished, so self-contained breathing apparatus and protective clothing must be worn when disconnecting piping, pulling blinds, and performing other operations preparatory to returning the unit to service.

At this point the shell of the heat exchanger being cleaned and the associated piping is rinsed from top to bottom at about 750 gpm with water until the pH of the water coming out is 1 unit or less different from that of the water going in. The unit is then ready for neutralization and passivation.

VIII.4.c Neutralization

Any residual acid in the shell of the heat exchanger can be neutralized by circulating a 2 percent solution of soda ash at 140°F for about 1 h, while agitating with nitrogen gas. The solution is prepared by transferring 400 lb of Na_2CO_3 to the supply tank, adding water, and agitating with steam until solution is complete. In some instances, the fresh water used for preparing the carbonate solution may contain enough dissolved oxygen to passivate steel, but in this procedure, in which steam is used to hasten the dissolution of soda ash and nitrogen is sparged into the loop to agitate the reirculating solution, there will be little, if any, oxygen left in the solution, so it is necessary to passivate the shell in a separate step.

VIII.4.d Passivation

It may be taken as a dictum that all carbon steel heat exchangers that are exposed to water or wet materials must be passivated before being returned to service after acid cleaning. Steel surfaces that have been exposed to acids are left in a highly reactive state, in which they corrode very rapidly. Passivators are inorganic oxidizing agents that are easily reduced, the most common being atmospheric oxygen and sodium nitrite. Zinc polyphosphate in aerated solutions forms a crystalline zinc phosphate identified by Trillat and Saison[15] as hopéite, $Zn_3(PO_4)_2 \cdot 4H_2O$, which dehydrates and becomes brittle at 375–400°F. These authors also identified phosphophyllite, $(Zn,Fe)_3(PO_4)_2 \cdot 4H_2O$, in surface films on steel. Passivation in the presence of phosphates affords two different types of corrosion protection: (1) phosphoric acid or alkali phosphates form a diffusion-barrier film of iron phosphates on steel surfaces; (2) if zinc is included, a hydrated, crystalline, protective film is formed. As might be expected, grease, oil, and other organic surface contamination prevent the formation of passive films. Brasses, bronzes, cupronickels, stainless steels, and monel are self-passivating, so this step is redundant when cleaning equipment constructed of these alloys.

The passivating solution proposed by Duncan, et al.,[16] containing 0.25 percent each of Na_2HPO_4, NaH_2PO_4, and $NaNO_2$ quickly forms a uniform passive film on steel. Before applying this solution, however, the contractor must ensure that no acidic residues remain in a vessel that can liberate the dangerous gas, NO_2. The passivation of the exchanger shell under consideration is accomplished by dissolving 50 lb of each salt in the supply tank, heating the resulting solution to 200°F, and circulating it through the shell for 15 min. The shell is allowed to soak for an additional 30 min, the solution is drained, and after rinsing briefly, the heat exchanger can be returned to service immediately.

As explained in Chapter V, pp. 110-112, when passivating steel equipment exposed to treated cooling water, it is convenient to passivate with a neutral chromate solution (1000 ppm CrO_4^{--}), as it can subsequently be flushed into the cooling system, thus avoiding a waste disposal problem. In nonchromate treated cooling systems, an aerated polyphosphate solution can be used instead.

VIII.5 TYPICAL WATER JETTING PROCEDURE

For most heat exchangers, particularly those with fixed tube sheets, high-pressure water jetting[17] is more economical than chemical cleaning, for in this method there is no hazardous waste produced to be disposed of, tubes can be cleared quickly, and there is little likelihood of damaging the equipment being cleaned. In the water jetting process, which is discussed further on pp. 98–102, water is forced through a nozzle containing several small orifices about 0.020–0.030 in. in diameter, at pressures from 2000 to 10,000 psi and at flow rates from 10 to 40 gpm. As a specific example, a nozzle containing 6 orifices 0.028″ in diameter can deliver water at 10,000 psi at a rate of 10 gpm. Focused jets of water at high pressure have a powerful cutting capability that enables them to clean metallic surfaces very efficiently; this makes the procedure especially convenient for clearing fouled heat exchanger tubes, as it is only necessary to remove the channel covers. If the exchanger is above ground level, boilermaker staging (named for the craft that customarily removes channel covers) is required. Under these conditions it is convenient to use flexible lances, which can usually be handled by a two-man crew, for water jetting the tubes. In case tubes are heavily fouled with hard deposits such as coke, or calcium salts, rigid lances must be used. As these may be up to 25 ft long, and require a three-man crew to force them through the plugged tubes, much more extensive staging is needed to enable the crew to walk the lance into and out of the exchanger tubes. When a polymer is added to the jetting water, as discussed on pp. 101–102, rigid lances are always used, regardless of the application.

Most chemical cleaning contractors have the necessary apparatus for water jetting industrial equipment, including a variety of lances, nozzles, pumps, and self-contained power units to supply the jet of high-pressure water. Although, as a rule, planned major shutdowns permit much greater productivity than jetting a single exchanger, a certain amount of unscheduled work must be requested on short notice, in which case contracts are normally let on a time-and-materials basis, rather than by firm bidding. Interested contractors then quote a rate per hour and make a rough estimate of the total cost of the job. If, however, the cost of a job that is not near completion approaches the estimate, a meeting should be convened to determine the reason and, if necessary, the contractor can be terminated. This discourages the making of unrealistically low estimates to secure jobs. In this type of agreement, the client usually pays no minimum, no setup time, no portal-to-portal pay, and no add-ons such as for lances, nozzles, hoses, and so forth.

To clean the outside of a tube bundle, it is necessary first to remove it from the shell, then transfer it to a specialized piece of equipment called an automatic bundle blaster. This hydraulically operated and remotely controlled device automatically rotates the tube bundle around its long axis, while at the same time a set of water jets traverses the length of the bundle, blasting deposits from the straps, tie-rods, spacers, and exterior of the tubes. These machines are useful only for square-pitched bundles; triangularly-pitched bundles must be either vat-dipped (an environmentally objectionable procedure) or chemically cleaned in place. Shells can sometimes be cleaned by a procedure called shot-gunning, in which a hand-held lance equipped with a nozzle and attached to a source of high-pressure water, is braced against the shoulder and actuated by a trigger. This is very tiring to the operator, and the results are erratic; few men can handle more than 10–12 gpm delivered at 4000–5000 psi.

To remove a deposit by water jetting, the impact pressure of the jet must exceed the bearing strength of the deposit, for which reason it is desirable to maximize the energy delivered per unit area. This can be done by increasing the water pressure, reducing the standoff distance (the distance between the tip of the nozzle and the surface of the deposit), increasing the size of the nozzle, or adding a polymer to the water to increase the cohesiveness of the jet. The throat of a nozzle is eroded rather quickly, producing turbulence in the stream of water and causing the jet to diverge; nozzles should be replaced frequently. Hard, dense, nonporous, sticky, or stringly deposits do not respond well to water jetting. Porous or laminar scale, however, is readily penetrated by the water jet, which exerts pressure that blasts the deposit from the underlying surface. A preliminary wash with hot water at about 150 psi often improves the results of high-pressure water jetting of waxy or oily tube bundles.

In U-tube bundles, both ends of each tube are rolled into a single tube sheet, most often in a vertical configuration with one leg above the other. They can be cleared by water jetting using a stainless steel flexible lance about 0.4″ in outside diameter; rubber tubing should not be used for this prupose. The flexible lance is first introduced into the upper leg with the water at reduced pressure (~ 2000 psi), which is then gradually increased as the tube clears. When the nozzle reaches the first bend, the lance is withdrawn and then introduced into the lower leg, where the procedure is repeated. No attempt should be made to force a flexible lance around a

U-bend; these tubes can almost always be cleared using the foregoing procedure.

Directions are now given for water jetting the tubes of an internal-floating-head heat exchanger without removing the floating-head cover. Fig. VIII.4 is a diagram showing the path of cooling water through such a heat exchanger having cooling water in the tubes. Only one tube in each pass is shown, although there are, of course, many tubes in industrial heat exchangers. The diagram depicts the exchanger with the channel cover removed (left end) and the floating-head cover in place (right end); the pass partitions that determine the path of cooling water through the device are also indicated. The technique for clearing this type of heat exchanger, which depends upon the continuous flushing of debris loosened by the water jetting from the floating-head chamber, is applicable to a variety of exchangers, condensers, and coolers. To carry out this procedure, it is first necessary to determine the position of the pass partition(s) behind the floating-head cover and to mark these locations on the tube sheet in the channel end of the heat exchanger so that the water-jetting crew can identify the bottom rows in each pass of tubes, as defined by the floating-head pass partition(s). Having made this identification, one tube in the bottom row of the first pass is jetted clear, and a flush lance (often called a dummy) without a nozzle is inserted into the tube. The flushing lance is attached to a source of water, which flows continuously into the floating-head chamber, flushing material dislodged by jetting out through the channel end of the heat exchanger. High-pressure water is not used for flushing, but a substantial flow ensures that all of the debris is washed out as jetting proceeds.

Because of the horizontal positioning of the pass partition in the floating-head chamber of the exchanger in Fig. VIII.4, all debris from tubes in the first and second passes collects on the bottom of the floating-head chamber, while that from the third and fourth passes collects on top of the pass partition. For this reason, the flushing lance may be inserted into a bottom tube of either the first or second pass to clear the bottom section of the floating-head chamber, and into a tube in the bottom row of either the third or fourth pass to clear the upper half of the chamber. Other configurations, of course, may have different requirements. When the flushing lance has been inserted and water is flowing, jet clear the rest of the tubes in the bottom row, then continue with the remaining tubes, working from the top row of the first pass downward. When finished with the first pass, rinse all of the tubes with water to ensure that they are clear. Remove the

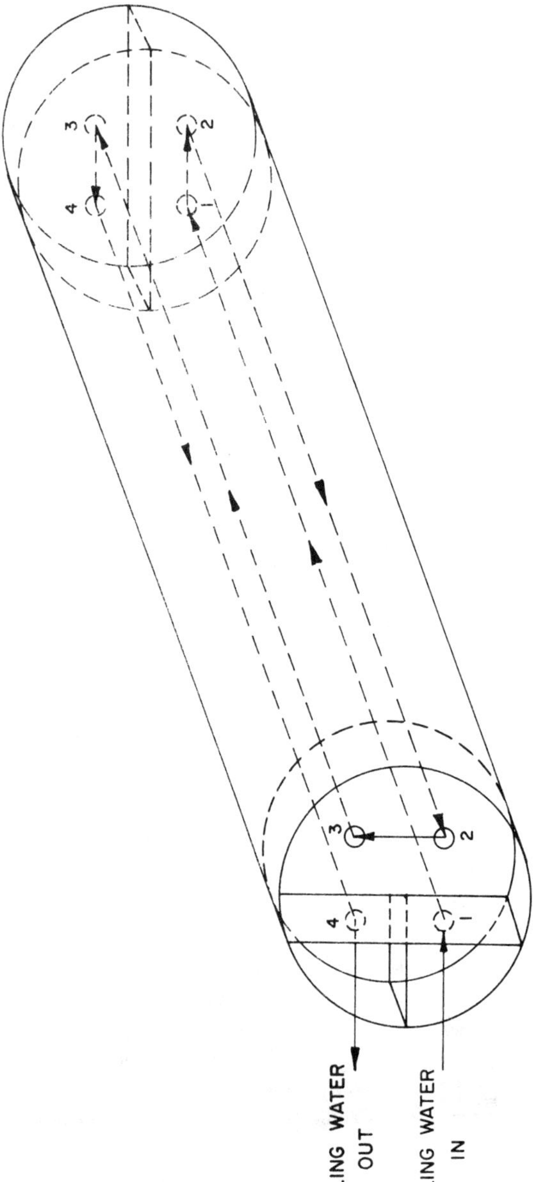

Fig. VIII.4 Four-pass heat exchanger with water in the tubes.

flushing lance and repeat the entire sequence in each of the remaining passes. After water jetting carbon steel exchangers they should be passivated as described in Section VIII.4.d, as the jetting procedure partially removes protective films formed by corrosion inhibitors.

After the tubes have been cleaned and passivated, the shell is hydrostatically tested by filling it with water and pressuring it against the blinds in the shell nozzles. If no leaks are discovered, replace the channel cover gasket and install the channel cover; if desired, the gasket can be tested by applying hydrostatic pressure through the tubes.

The savings that can be realized by using the foregoing procedure can be significant. Labor costs to remove and replace the shell and floating-head covers are avoided, while costs of staging, bolt cleaning, and crane time are much reduced. When this technique is used, however, it is impossible to inspect the shell or collect samples, except through shell nozzles or drains. Also, it is worthwhile to remove and inspect the floating-head cover in exchangers in which corrosion of the floating-head or failures of the floating-head gasket have been experienced. If, as it usually is, time is an important consideration, water jetting can probably be accomplished faster if the floating-head cover is first removed.

CAUTION: The accident potential is high in any high-pressure water jetting procedure.

References

(1) McCoy, J. W. 1974. *The chemical treatment of cooling water.* New York: Chemical Publishing Co.

(2) NACE Standard RP-01-70. 1970. Protection of austenitic stainless steel in refineries against stress corrosion cracking by use of neutralizing solutions during shut down.

(3) Perry, J. H. 1963. *Chemical engineers' handbook.* 4th ed. New York: McGraw-Hill Book Co.

(4) Plunkett, E. R. 1976. *Handbook of industrial toxicology.* New York: Chemical Publishing Co.

(5) Tewari, P. H. and Campbell, A. B. 1976. Dissolution of iron sulfide (troilite) in aqueous sulfuric acid. *J. Phys. Chem.* 80 (17): 1844.

(6) Lawson, M. B., Martin, L. D., and Arnold, D. G. 1980. Chemical cleaning of FeS scales. Paper No. 219. *Corrosion/80.* NACE. Chicago.

(7) McCoy, J. W. 1969. *Chemical analysis of industrial water.* New York: Chemical Publishing Co.
(8) Westcott, C. C. 1978. *pH measurements.* New York: Academic Press.
(9) Deane, J. A. 1974. *Lange's handbook of chemistry.* 11th ed. New York: McGraw-Hill Book Co.
(10) *Gas Conditioning Fact Book.* 1962. Midland, Michigan: The Dow Chemical Co.
(11) Uhlig, H. H. 1967. *Corrosion and corrosion control.* New York: John Wiley & Sons.
(12) Sax, N. I. 1979. *Dangerous properties of industrial materials.* 5th ed. New York: Van Nostrand Reinhold Co.
(13) Elkins, H. B. and Fahy, J. P. 1967. Arsine poisoning from aluminum tank cleaning. *Ind. Med. & Surg.* 36: 747.
(14) Stander, R. A. Mar. 17-21, 1958. Precautionary procedures in chemical cleaning. NACE Publication 8A159. Fourteenth Annual Conf., Nat'l. Assoc. of Corrosion Eng., San Francisco, California.
(15) Trillat, J. J. et Saison, J. 1961. Etude des couches de phosphation au moyen de rayons X et de la diffraction électronique, *Mem. Sci. Rev. Met.* 58: 496.
(16) Duncan, F. R., Fyock, D. H., Baldwin, L. V., and Hopkins, R. D. 1972. Preoperational cleaning of high pressure boilers of General Public Utilities Corporation. *Proc. Int'l. Water Conf., Eng. Soc. W. Pa.* 33: 97.
(17) Tursi, T. P., Jr. and De Lecce, R. J., Jr. 1976. Development of very high pressure water-jet for cleaning Naval boiler tubes. *Marine Tech.* 13 (3): 272.

IX

Columns, Reactors, and Other Vertical Vessels

In the preceding two chapters, procedures have been described for chemically cleaning tubed equipment. Now it is appropriate to discuss the cleaning of various types of vessels used in petroleum refining and chemical processing that are essentially vertically oriented shells with or without certain internal fittings. These include reactors, strippers, rectifiers, fractionators, and reboilers in which physical separations are made that depend upon differences in boiling points, or in which chemical conversions such as polymerization, condensation, isomerization, or hydrogenation are accomplished. In Section IX.1 the functions of various columns and the types of fouling that they may contain are described; Section IX.2 contains a discussion of various cleaning operations; and in Section IX.3 a chemical cleaning procedure generally applicable to columnar equipment is given.

IX.1 TYPES AND FUNCTIONS OF INDUSTRIAL COLUMNS

The most widely used columns in industry are those in which mixtures of liquids are separated into specific boiling ranges. Some of the more familiar distillation columns are used to fractionate crude petroleum, first at atmospheric pressure, then at a reduced pressure of 50–100 mm Hg. Atmospheric columns vary in dimensions and internal fittings, one of the larger sizes is about 14 ft in diameter and 150 ft in height. Within the column, trays are mounted one above another at regular intervals approximately 25 in. apart, to form a multiplate fractionator. The type of column under discussion contains 50-55 trays, each pierced with 10-30 rows of holes fitted with chimneys, and covered by movable bubble caps, as shown in Fig. IX.1.

Crude petroleum is preheated by exchange with hot side streams from the distillation column, then it is passed through tubes in a fired furnace, where it reaches a temperature of 650-700°F. The hot feed usually enters the column near the fifth or sixth tray from the bottom in a region known as the flash zone. Vaporized lower boiling components rise upward through the trays in the rectification zone, and the higher boiling components move downward toward the bottom of the column. The lower four trays are in the stripping section, so called because steam is injected here, at the rate of 5-15 lb per barrel of feed, to strip entrained lighter components from the reduced crude before it leaves the atmospheric column and is pumped into the vacuum column.

As shown in Fig. IX.1, hot vapor rises through the bubble caps on each tray, passing through a liquid layer supplied by downcomers from trays above. Liquid moves across each tray under the influence of a hydraulic gradient, then flows over an adjustable weir, and through a downcomer to the tray below. On each tray, higher boiling components condense, release their latent heat of vaporization, and vaporize lower boiling components in the liquid, which then rise as a vapor to the tray above. This process constitutes a mass transfer of the liquid phase to the vapor phase. Gases and light gasoline (80-150°F) are taken from the top of the column; other products including naphtha (150-315°F), kerosene (315-490°F), automobile diesel (490-650°F), and light gas oil (650-700°F) are removed at appropriate levels from the column through side cut draws. These temperature ranges are approximate; the actual side cuts that are obtained depend upon the crude slate available, product demand, and requirements of other integrated processes.

Crude petroleum distillation columns are usually made of carbon steel; bubble caps may be cast iron, pressed steel, or alloy steel. Austenitic stainless steels are frequently selected for equipment used in the chemical processing industry, as they are resistant to a great many chemicals. Nickel-based alloys including duranickel, inconel, and monel are used for certain equipment in hydrofluoric acid alkylation plants. In one of the many applications of the alkylation process, isobutane is condensed with isobutene using HF as a catalyst. Anhydrous HF and the mixed isomers are introduced into steel contactors, after which HF is separated by settling, then concentrated for reuse in a regenerator. The hydrocarbon layer, con-

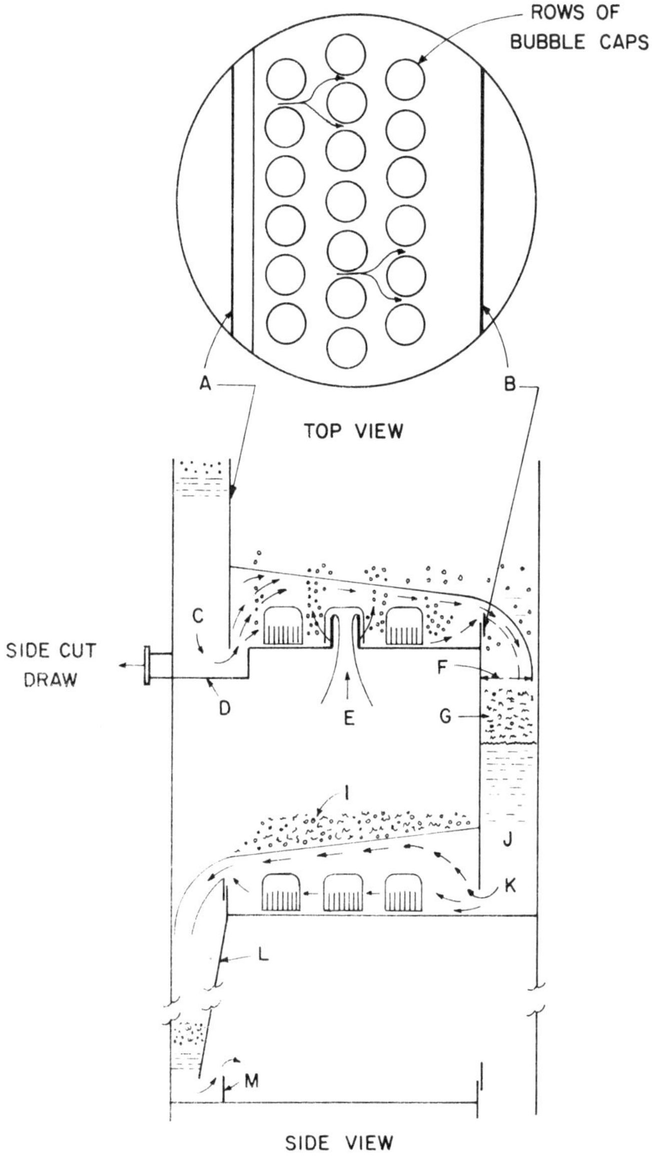

Fig. IX.1 Bubble cap tray column. A, Straight downcomer; B, Adjustable outlet weir; C, Liquid; D, Recessed seal pan; E, Vapor and mist; F, Throw over weir; G, Froth; H, Disengaging space; I, Foam, froth, and bubbles; J, Clear liquid; K, No seal; L, Tapered downcomer; M, Inlet weir.

taining alkylate with small amounts of HF and water, is transferred to a steel recycle rectifier column containing 30–40 bubble-cap trays, and equipped with a water-cooled overhead condenser constructed of monel. Butanes and anhydrous HF move up the column, while the HF-water azeotrope (35.6% HF-64.4% H_2O) and alkylate move down the column in the first step in purifying the alkylate product.

At various stages in the processing of petroleum, steam is injected to cool reactors and prevent coking, for atomizing hydrocarbons through nozzles, or to absorb excess hydrogen sulfide and ammonia formed in several catalytic processes. Condensed steam separates in knockout pots and reflux accumulators, and is then stripped of hydrogen sulfide and ammonia. The foul water is heated by steam in a preheater to about 200°F, after which it passes into an H_2S stripper consisting of a short column about 2 ft in diameter containing five or six sieve or shower trays. Stripping steam at 240°F is admitted near the bottom of the column, and H_2S and NH_3 pass out overhead while stripped water is withdrawn from the bottom of the stripper. Both preheater and stripper are likely to be fabricated of carbon steel. Sieve or perforated shower trays consist of plates in which holes have been drilled. As the liquid on these trays is supported by vapor from below, columns containing sieve trays do not operate well at low rates.

Packed columns contain no trays or plates, but instead are filled with inert packing such as Raschig rings, saddles, or helices made of ceramic or polyethylene. In this arrangement, which is most suitable for absorption processes involving fluids or solutions that are prone to foam, thin films of refluxing liquid are distributed over the large surface area of the packing throughout the column. Because of the tendency of refluxing liquid to flow down the walls of larger columns, the diameter of a packed column is usually limited to about 2 ft. For the same reason, tray columns are preferable for separations requiring many theoretical plates, even though the hold-up of liquid is greater in tray columns than in packed columns. Owing to the relatively small size of these vessels, it is often feasible to clean them with solvents without removing the packing.

IX.2 THE NATURE OF COLUMN FOULING

Fouling in fractionating columns used to distill crude petroleum tends to be heaviest in the flash zone, normally between the fourth and the seventh tray from the bottom of the column. Deposits here may consist of mixtures of organic and inorganic materials, mostly Fe_2O_3, Fe_3O_4,

FeS, FeS_2, and, occasionally, $Fe(FeO_2)_2$, ferrous ferrite, enclosed in polymerization products, cokes, and bituminous material. Ordinarily, organic material predominates, 50–80 percent, with 20–50 percent of inorganic oxides and sulfides. In order to conserve energy, crude petroleum feed is preheated in one or more feed-effluent heat exchangers in which cold crude oil is passed through the tubes while a hot product, e.g., the overhead stream from the vacuum column, is passed through the shell. In this process the feed is heated to about 300°F. This reduces its viscosity and inhibits the formation of oil–water emulsions in the next processing step, which is desalting. Salts introduced into petroleum in underground brines, sand, silt, silicates, drilling muds, and corrosion products cause plugging, coking, and further corrosion in the furnace tubes and in the distillation column. Magnesium salts and hydrogen sulfide are highly corrosive to iron, especially in the overhead piping and condenser on the atmospheric column. Many crudes contain H_2S, and most others that contain sulfur compounds such as thiophene, C_4H_4S, and its derivatives, release H_2S upon heating, which rapidly reacts with iron to form FeS.

$$Fe + H_2S = FeS + H_2 \quad \text{(IX-1)}$$

Magnesium chloride, when heated in the presence of small amounts of water, hydrolyzes releasing HCl.

$$MgCl_2 + 2H_2O = Mg(OH)_2 + 2HCl \quad \text{(IX-2)}$$

FeS initially forms a hard corrosion barrier on iron surfaces, but HCl formed by the reaction in Eq. (IX-2) dissolves this barrier, exposing fresh metal to the action of both H_2S and hydrogen ion.

$$FeS + 2H^+ = Fe^{++} + H_2 \quad \text{(IX-3)}$$

Because most of any magnesium salts present in crude oil are extracted in the desalting step, H_2S normally predominates, and large amounts of FeS are formed, both in the furnace and in the distillation column.

To prevent the entrainment of asphalt in vacuum column side cuts, a demister pad of woven wire alloy gauze, 6-12 in. thick, is installed just above the flash zone. This is continuously cleaned by injecting wash oil, but eventually the interior of the pad plugs with coke and when this happens the pad must be removed and discarded, for there is no effective way to clean a plugged demister pad.

Gypsum, $CaSO_4 \cdot 2H_2O$, is formed both in desalters and in preheat exchangers in petroleum processing. If dehydrated below 570°F, a soluble anhydrite is formed that rapidly takes up water; at higher temperatures a very slightly soluble anhydrite forms that eventually yields to 3–4 percent HCl at 200°F. Sodium gluconate solution may be superior to HCl for dissolving any of the calcium sulfates, however, for it rapidly dissolves milk stone, the very hard, insoluble calcium caseinate, that is so troublesome to remove from dairy equipment.

Columns that are used for separating relatively clean products such as alcohols, xylenes, glycols, and the like, tend to be neglected for long periods of time because they cause no apparent trouble. In such columns thick layers of rust are often formed in which the outer layers are quite soluble in acids, but the inner layers are singularly insoluble. This has to do with aging and the variations in wettability of different forms of iron oxide. Then too, as the upper, center, and lower sections of a column are generally exposed to different temperature gradients and particular compositions of processed materials, each section may contain a unique type of scale with a solubility differing from those in the other sections. Even though a representative sample can be obtained and proves to be readily soluble in the laboratory, chemical cleaning will be unsuccessful if holes in sieve trays are blocked, bubble caps or chimneys are fouled, or if any other mechanical occlusion prevents the circulation of cleaning solution.

The sampling and storage of material from hydrofluoric acid systems presents special difficulties. In large plants it may be worthwhile to have a monel can which can be fabricated for sampling, although these are expensive. Samples of this type should be clearly marked, preferably with distinctively colored tags, and handled with appropriate caution. Corrosive systems are generally washed with water before being opened for inspection and maintenance and this, of course, inevitably alters the composition of any deposits that are present. The deposition of acid salts, e.g., KHF_2, is common in some sections of hydrofluoric acid alkylation plants, but they are ordinarily dissolved in the preliminary washing operation. Corrosion products from the regenerator (monel) tend to migrate into the recycle rectifier column and accumulate on the trays in the form of granular mud. In addition, the overhead condenser on this column is subject to perforation by corrosion on the water side, which can lead to the formation of extensive corrosion deposits within the column. Significant quantities of fluoride salts of iron, copper, and nickel, along with HF are found in these deposits.

Although hydrogen sulfide stripper columns are customarily fabricated of carbon steel, the water that is stripped in them usually will have been in contact with overhead condensers and tube bundles of monel, admiralty, and various copper-nickel alloys. For this reason, deposits containing cuprous sulfide, Cu_2S, ferrous sulfide, FeS, and nickelous sulfide, NiS, are often encountered in preheaters and on trays in the stripping column, with cuprous sulfide predominating. In general, concentrations of FeS are low on stripper trays, indicating that little or no corrosion occurs in that vessel and suggesting that all of the metallic sulfides deposited in the unit are carried into it in the water as corrosion products of copper–nickel–iron alloys ahead of the stripper. Monel, for instance, is not notably resistant to sour water. Most of the deposition of iron and nickel occurs in the preheater because of the abrupt rise in temperature there and because these two sulfides are less soluble in ammonia than is cuprous sulfide. All of the sulfides are probably in colloidal solution when they enter the preheater, where they are coagulated by the sudden increase in temperature.

Methods for chemically analyzing all of the deposits discussed in this section can be found in another publication.[1]

IX.3 SELECTION OF SOLVENTS

Without exception, fouled crude oil distillation columns must be deoiled before making any attempt to dissolve the organic portion of the deposit. Residual asphalts and tars, which often foul these units, contain condensed aromatic rings that are highly soluble in benzene and benzene derivatives. Therefore, if provision can be made for reclaiming an organic solvent, as described on p. 40, ortho-dichlorobenzene, a chlorinated aromatic hydrocarbon not previously mentioned, is the solvent of choice for removing asphaltic residues. Hollingsworth, et al.[2] have reported on the toxicity of o-dichlorobenzene, which is a central nervous system depressant that damages the liver and kidneys. According to Sax,[3] the ortho isomer is probably more toxic than either the meta or para compounds. Among the symptoms produced by exposure to the solvent are headache, dizziness, weakness, nausea, swelling of the hands, feet, and ankles, and jaundice.[4]

Once a column has been deoiled, most of the inorganic material can be dissolved or, at least, disintegrated by dilute inhibited hydrochloric acid. Attention must be paid, however, to the metallurgical composition of the column; some of them contain stainless steel trays or bubble caps, although crude oil fractionators are usually made of carbon steel with cast iron bubble caps. Also, as ferrous sulfide is nearly always present, provision must be made for absorbing or flaring H_2S, as described on pp. 165-171, Sometimes the evolution of H_2S can be minimized or eliminated entirely by resorting to an alkaline transformation of FeS, as suggested by Lawson, et al,[5] for disintegrating iron sulfide deposits.

TABLE IX-1

Stability of Iron Compounds

Compound	*Solubility product*	*Formation constant*
FeS	6.3×10^{-18}	1.6×10^{17}
$Fe(OH)_2$	8.0×10^{-16}	1.3×10^{15}
$Fe(OH)_3$	4.0×10^{-38}	2.5×10^{37}

A consideration of the values in Table IX-1 suggests that FeS would not be affected by a high concentration of hydroxyl ion alone, but that in an aerated alkaline solution the very small value of the solubility product of ferric hydroxide would tend to force the reaction represented by Eq. (IX-4) to the right.

$$2FeS + 3H_2O + 3/2O_2 = 2Fe(OH)_3 + 2S \qquad \text{(IX-4)}$$

Bell[6] has formulated an exchange reaction to predict the effect of ligands, in particular chelants, on slightly soluble compounds. In the example under discussion,

$$FeS + 2OH^- = Fe(OH)_2 + S^{--} \qquad \text{(IX-5)}$$

the exchange constant, K_e, is calculated from the formation constant of FeS and the solubility product of $Fe(OH)_2$.

$$K_e = 1/(K_f \times K_{sp}) \qquad \text{(IX-6)}$$

If $K_e > 1$, i.e., if $(K_f \times K_{sp}) < 1$, then FeS can be transformed to $Fe(OH)_2$ by hydroxyl ion. Substituting the appropriate values from Table IX-1,

$$K_e = 1/(1.6 \times 10^{17} \times 8.0 \times 10^{-16})$$
$$= 1/128$$

showing that FeS is stable in the presence of alkali. If, however, an oxidizing agent is present to oxidize ferrous to ferric ion, as in Eq. (IX-4),

$$K_e = 1/(1.6 \times 10^{17} \times 4.0 \times 10^{-38}) = 4.2 \times 10^{20}$$

indicating that under this condition FeS is readily transformed to $Fe(OH)_3$. The rate of the reaction represented by Eq. (IX-4) can be speeded up considerably by including sodium gluconate in the sodium hydroxide solution, as the dissociation constant (stability constant) of ferric gluconate is 6.3×10^{-38} See p. 76 for the composition of a suitable formulation containing gluconate, and the method of applying it; an acid wash is usually required after the alkaline treatment to remove hydrous oxides, iron oxides, and any residual FeS. As pointed out before, alkaline scrubbers must always be used when acidifying deposits that may contain sulfides.

Kowalski(7) has described a blend of the chelants nitrilotriacetic acid, citric acid, and 1-hydroxyethylidene-1,1-diphosphonic acid,* which he found to be highly effective for dissolving calcium carbonate, cupric oxide, and magnetite. As this combination of ingredients is reported to yield exceptionally clean surfaces when applied to heavily rusted equipment,** it should be of special value in cleaning columns containing aged rust that is quite resistant to hydrochloric acid as noted in Section IX.2, above. A suitable blend is as follows:

DEQUEST® 2010	1.1 percent	(60% active)
Citric acid	1.9 percent	(50% active)
Trisodium nitrilotriacetate	1.5 percent	(93.5% active)

The pH of the prepared blend should be adjusted to 5-7, preferably with sodium hydroxide and phosphoric acid, then the solution is applied for 2-3 h at 170°F, while agitating with nitrogen.

Columns in HF alkylation plants are normally rather clean with respect to carbon compounds, but the removal of the inorganic fouling described in Section IX.2, above, which often can be done by water jetting, may be complicated by the presence of hydrofluoric acid. Although this acid causes extremely dangerous and painful burns,(8) it can be rendered

* Manufactured and sold under the trade name DEQUEST® 2010 by Monsanto Industrial Chemicals Company, 800 N. Lindbergh Blvd., St. Louis, Missouri 63166.

** Karl Hirsch, Miller Brewing Company, Milwaukee, Wisconsin. Private communication.

innocuous by a preliminary flush of the column with a dilute solution of boric acid, which readily reacts with HF to form fluoboric acid.

$$H_3BO_3 + 4HF = HBF_4 + 3H_2O \qquad (IX\text{-}7)$$

The latter is completely hydrolyzed in alkaline solutions.

$$BF_4^- + 4OH^- = H_2BO_3^- + 4F^- + H_2O \qquad (IX\text{-}8)$$

The compound $Fe_2F_4 \cdot 7H_2O$ is also sometimes found deposited on the overhead condensers of depropanizer or recycle rectifier columns in these plants.

Desalter sediments often contain arsenic, a common component of crude petroleum, as do absorbers, reclaimers, coolers, and associated piping in alkanolamine gas sweetening units in which sodium arsenite is, or has been, used as a corrosion inhibitor. The principal hazard is the release of arsine, AsH_3, during acid cleaning; several arsenic compounds that have been found in deposits from monoethanolamine systems are mentioned on p. 172.

The chemical cleaning of hydrocracking reactors is seldom required, but contractors should be aware of a unique hazard that exists when handling catalysts that contain nickel. Nonregenerated catalyst, which is pyrophoric, may be unloaded from a reactor into barrels either wet or dry. If it is removed dry, it is customary to maintain a nitrogen blanket on the catalyst bed as it cools down from about 400°F. If, instead, it is allowed to cool in an atmosphere containing carbon monoxide, the highly toxic and volatile nickel carbonyl, $Ni(CO)_4$, is formed. Testing the vapor space of a barrel of catalyst for this compound can be misleading in that the vapor space may be free of $Ni(CO)_4$, yet if the catalyst is agitated slightly, a considerable quantity of the toxic vapor may be volatilized. Plunkett[4] gives the threshold limit value for $Ni(CO)_4$ as 0.001 ppm. Inhalation of the vapor damages the respiratory tract, central nervous system, liver, adrenals, and kidneys. Chronic exposure leads to carcinoma of the nasal sinuses and lungs. In fatal poisonings, death is usually caused by pulmonary and cerebral edema.[9]

IX.4 METHODS OF CLEANING COLUMNS

As in all chemical cleaning, the most effective solvent for removing fouling from a column can only be determined by solubility tests on representative samples from the column. Because of differences in temperature and chemical characteristics of materials being processed, it is commonly found that the compositions and solubilities of samples of fouling from different sections of a column are highly variable. Thus, the lower section of a crude distillation column, in which high-boiling petroleum fractions are concentrated, tends to foul with tars and polymeric materials that are best dissolved by a solution containing a nonionic wetting agent and an aromatic solvent. The middle and upper sections, on the other hand, usually contain inorganic scales and sediments consisting largely of oxides and sulfides of iron. Voluminous corrosion products, mostly iron oxides, are often concentrated on trays where water condenses and is withdrawn. Such inorganic accumulations are usually susceptible to dilute inhibited hydrochloric acid. These differences in composition of fouling must be taken into consideration when sampling and determining solubilities, as a sample taken from the top section of a fouled column may suggest a solvent that will give decidedly inferior results in the lower section of the same column. Successful chemical cleaning of distillation columns cannot be assured unless samples of fouling material can be obtained for solubility testing. Chemical cleaning of columns is, in general, most efficient if done routinely to prevent the formation of rust or sulfide scales that block bubble cap chimneys or the perforations in sieve trays. It is rarely effective in any unit in which the passages are obstructed to the point of operational failure, although when followed by mechanical cleaning, acid cleaning is sometimes the first step in restoring such a column for service.

Although not a chemical cleaning procedure in the usual sense, it is worthwhile mentioning, in passing, the use of a mixture of alkali and detergent for sweetening, or safeing, a desulfurizer or hydrogen sulfide stripping column so that it can be entered safely for inspection or mechanical repairs. This can be accomplished in as short a time as 8 h by introducing a hot (180°F) solution containing 1 percent of NaOH and 0.1 percent of a surfactant into the top of the column and allowing it to cascade down through the trays to the bottom, and then recirculating the solution back to the top with a suitable pump. The alkaline solution neutralizes hydrogen sulfide and washes loose scale out of the column, permitting a safe entry and, incidentally, providing a cleaner working area than would otherwise be encountered. The older procedure of steaming and water washing often required up to 6 days to strip a column of H_2S and make it safe for entry.

High pressure water jetting is occasionally used to remove friable deposits that tend to shatter or chip, but it is a hazardous, dirty, tiring procedure that when applied to fractionating columns requires 2 to 3 days to complete. Hydrogen sulfide, if present, must be removed as described above before workmen can enter the column through tray manways. Debris is water jetted off the trays, one at a time, and flushed down the column. Although the short length of the hand-held water gun, the cramped working space (24–30 in. between trays), and the internal fittings in these columns make this procedure unusually hazardous, it is the method of choice for removing coke deposits; all others are best removed by chemical cleaning.

IX.4.a Flooding

Filling a column with an appropriate cleaning solution and then recirculating it from the bottom over the top is the most effective way to clean columns containing bubble cap, sieve, raindeck, or valve trays, for this exposes all parts of the vessel, including the bottoms of the trays, to the cleaning solution. Before filling a column, however, it is essential to ascertain whether the vessel is capable of bearing the weight of the volume of solution to be used. The supporting skirts of columns intended for vapor service, for instance, are not designed to support very much extra weight, and the skirt is likely to collapse if such a vessel is filled with water or cleaning solution. The approximate capacity of a column can be calculated from its height and diameter, with about 20 percent subtracted to account for the volume occupied by the trays, bubble caps, weirs, and other internal fittings. For example, a column 14 ft in diameter and 150 ft high, if empty, would have a capacity of,

$$\pi(14)^2 \times 150 \times 7.48/4 \cong 173{,}000 \text{ gal}$$

Correcting this value by subtracting 20 percent gives,

$$173{,}000(1-0.2) \cong 138{,}000 \text{ gal}$$

If the column were filled with 10 percent HCl, specific gravity 1.047, the total weight of the solution would be,

$$138{,}000 \times 1.047 \times 8.34 \cong 1{,}200{,}000 \text{ lb}$$

As it would be undesirable to subject any vessel to this much additional weight, even if the expense of the solution and its disposal were acceptable, large columns are usually cleaned by the more economical procedure outlined in Section IX.4.b, below.

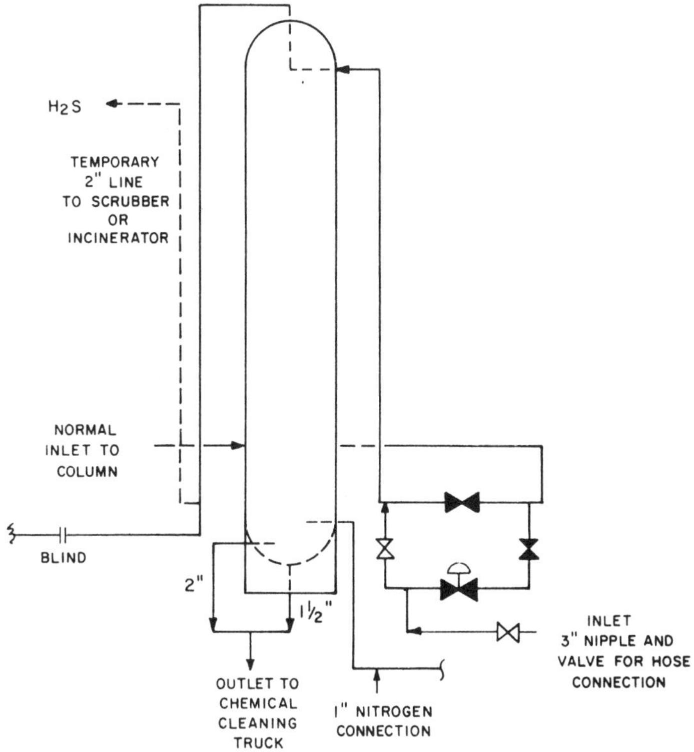

Fig. IX.2 Connections required for cleaning a column.

Before commencing a chemical cleaning, it is first necessary to select inlet and outlet connections to be used for circulating the cleaning solution, after which blinds are installed to isolate the column and to ensure that no lines are open to the atmosphere, drains, or nondrainable deadlegs, and that steam and water services to the column cannot be inadvertently contaminated. When these steps have been completed the column is filled through the lower connection, marked A in Fig. IX.2, until the solution flows from the top of the column and returns to the chemical cleaning truck through B. The direction of flow is then reversed, suction being taken by means of a reversing manifold through A, and the solution being pumped to the top of the column through B. During this operation the vent, E, on the top of the column remains closed. Because the ratio of acid to deposits is large in this flooding procedure, the spending rate of the acid cannot be depended upon to indicate when the column is clean. Instead, a minimum time of 6–8 h should be prescribed in addition to determining the acid concentration periodically.

Following the acid treatment the column is rinsed through B with water, the pH being checked occasionally to indicate when rinsing is complete, after which the unit is filled with a neutralizing and passivating solution, circulated for 1 h, then drained. When draining a column it is important that a vent be opened at the top before opening the drain line, to avoid the possibility of collapsing the vessel when the pressure inside of it is reduced by the weight of the draining solution. In instances where H_2S has been generated during cleaning, 0.5 percent of NaOH should be added to the neutralizing solution if the column is subsequently to be entered for inspection or repair.

IX.4.b Cascading

When flooding is impractical because of the large weight and volume of solvents required, cascading a cleaning solution through a column is a suitable procedure for those containing bubble cap or valve trays. Cascading is less effective for partially obstructed sieve or raindeck trays, however, as the solvent preferentially flows through open passages with little if any effect on those that are blocked. Other things being equal, the choice between flooding and cascading depends upon the size of the column to be cleaned and the economic aspects of the use and disposal of the necessary solvents. When specifying either method it is essential, as always, to take into account the metallurgical composition of the equipment to be cleaned. Bubble caps, for instance, are often made of cast iron, which limits the temperature of acidic solutions to a maximum of 140°F.

A column is cascaded by filling it through B to a calculated depth C or D, shown in Fig. IX.2, then taking suction from A, and circulating the cleaning solution through B at the maximum pumping rate of the contractor's equipment, say, 450 gpm minimum. This causes the solution to rain through the column, filling the trays, splashing down over each succeeding tray, and collecting in the bottom. A stream of nitrogen gas injected into the bottom of the column in the vicinity of A is useful for agitating the solution on the trays.

Filling the column to the level indicated by C provides the minimum amount of solution necessary to maintain sufficient working head pressure on the circulating pump. To achieve satisfactory pumping, the line to the pump suction should be larger than that on the discharge. Thus, when pumping into a 2-in. inlet line it is advisable to provide two 3-in. suction lines teed off of an existing 4- to 6-in. drain or bottoms line from the column. Should the supply to the suction be insufficient, the pump will cavitate and be unable to discharge its designed capacity. Experienced chemical cleaning contractors know what is best suited to their own equipment and carry the required flanged pipe fittings for making proper connections.

D indicates the level to which one might wish to fill a column in order to flood, and thus clean more efficiently, certain trays. In most distillation columns the lower trays are exposed to more heat and higher boiling materials than are the upper trays; consequently, they foul faster and more severely than do the upper trays. It is for this reason desirable to establish a level of solution in the column that keeps the lower trays submersed, thereby providing a more rigorous cleaning than is required for the upper trays. In this procedure the lower trays are flooded while the upper trays are cascaded. The volume of solution needed to fill a column to C or D can be calculated from the formula,

$$V, \text{gal} = h\pi d^2 \times 1.5 \qquad \text{(IX-9)}$$

where:

h = desired height of solution, in feet
d = diameter of the column, in feet

Alternatively, a pressure gauge installed in the drain line can be used to fix the level by noting that,

$$P, \text{psi} \cong h/2.3 \qquad \text{(IX-10)}$$

The level is then raised or lowered until the gauge indicates the pressure corresponding to the desired height, h. When cascading columns fitted with bubble cap or valve trays, more solution is required to fill to levels C or D than is calculated from Eq. (IX-9), because of hold up on the upper trays (see Fig. IX.1). The volume held up can be approximately calculated from the diameter of the column, the area of the downcomers, and the crest of the weirs.

Suppose:

the diameter of the column is 14 ft
the area of the downcomers is 30 ft^2
the crest of the weirs is 2.5 in.

Then the hold up on each tray is,

$$[\pi(7)^2 - 30] \times 7.5 \times 2.5/12 \cong 194 \text{ gal/tray}$$

If, for example there were 20 trays above D, it would be necessary to add 20 × 194 = 3880 gal of solution in addition to the volume calculated using Eq. (IX-9).

IX.4.c Foaming

Foamed acid has proven to be an efficient medium for chemically cleaning fouled surface condensers and other large industrial vessels that require excessive amounts of solvent to fill them. By substituting foam for a liquid cleaning solution, stress on the vessel being cleaned is much reduced, as is the expense of purchasing and disposing of solvent. Furthermore, foam transports dislodged insoluble material out of a vessel better than does a liquid, it is itself easy to displace after use, and it is readily produced as required in patented foam generators at the cleaning site. The modes of operation of a number of generators have been disclosed.[10-13] In most of these, foam is generated by sparging nitrogen—in some applications, air—through a gas dispersion tube having openings about 0.004 in. in diameter, into a prepared solution of cleaning solvent confined in the apparatus. The outlet of the generator is connected to the vessel to be cleaned, into which the generated foam flows continuously, impelled by the pressure of the gas.

As the rate of formation of foam must be substantially greater than its rate of collapse in order to fill a vessel completely, a surfactant must be added to aid in stabilizing the foam. Brenner and Engle[10] have disclosed the results of tests using three anionic surfactants: sodium N-methyl-N-oleyltaurate, sodium cocamine acetate, and ammonium isopropylbenzenesulfonate. They also obtained good results with a nonionic polyglycol prepared by condensing 1 mole of disecondarybutylphenol with 10 moles of ethylene glycol. All of these surfactants produced foam of adequate stability when added to liquid formulations in concentrations ranging from 0.1 to 0.5 percent by volume. Carroll and Rodgers[11] obtained satisfactory foam using sodium N-lauroyl-β-iminodipropionate at a concentration of 0.5 percent by volume.

Crowe and Duranleau[12] have disclosed the use of dichlorotetrafluoroethane, $CClF_2CClF_2$, Freon 114®,* as a foaming agent. In their procedure, foaming agent, stabilizer, and inhibited concentrated solvent are introduced continuously into dilution water heated to 150-200°F. Freon 114® vaporizes, forcing the foam generated through the vessel to be cleaned; upon emerging, the foam is externally cooled, condensing the Freon 114® and collapsing the foam, which is then recycled. The stabilizer suggested for use in this method is methyldiethylperfluorooctyl sulfonamide quaternary ammonium iodide,

$$\left[\begin{array}{c} \qquad\qquad\qquad\qquad\qquad\quad C_2H_5 \\ \qquad\qquad\qquad\qquad\qquad\quad | \\ C_8F_{17}SO_2NHCH_2CH_2{-}N{-}C_2H_5 \\ \qquad\qquad\qquad\qquad\qquad\quad | \\ \qquad\qquad\qquad\qquad\qquad\quad CH_3 \end{array}\right]^{+} I^{-}$$

a salt not easily hydrolyzed. Zimmerman and Lavine(14) report that the boiling point of Freon 114® is 38.4°F, and that it exerts a pressure of 12-69 psi at 70°F.

The quality of foam is judged by its wetness, which is the ratio of the volume of foam to the volume of liquid condensed when the foam collapses. This volume ratio is determined by the rate of gas flow to the generator: the higher the gas flow, the lower the volume ratio, i.e., the wetter the foam. Wet foam cleans and dissolves scale faster than dry foam, the optimum volume ratio being about 20:1. Brenner and Engle(10) suggest that foam be generated at such a rate that its linear velocity over the internal surfaces of a vessel being cleaned is at least 0.1 ft/min, and that its collapse rate is such that at least 0.01 ml of liquid per minute per liter of foam is released to wet the surfaces.

In the presence of hydrocarbons, foam collapses immediately when prepared by the procedures thus far described. Brenner and Engle(10) found, however, that kerosene, diesel fuel, and trichloroethylene could all be foamed satisfactorily if a saponified mixture of tall oil and caprylic acid was included with the usual surfactants. Tall oil is a mixture of abietic (see p. 68 for formula), oleic, and linoleic acids, with lesser amounts of a few other unsaturated fatty acids.(15)

The following prescribed sequence of steps was used to remove an accumulation of mixed iron oxides from a column 22 ft high and 3 ft in

---

* Manufactured by E. I. Du Pont de Nemours, Petrochemicals Department, Wilmington, Delaware 19898.

diameter fitted with 6 trays, each containing 16 bubble caps. The column was filled with foamed inhibited hydrochloric acid containing an abietylamine corrosion inhibitor and a surfactant. Filling was accomplished in approximately 10 min, after which the acidic foam was recirculated for 3 h. Upon allowing the foam to collapse, 90 gal of liquid condensed from the 1200 gal of foam required to fill the column. Thus, cleaning was successful with 7.5 percent of the volume of hydrochloric acid that would have been needed to flood the column.

Subsequent rinsing, neutralizing, and passivating steps are all done with foam, each succeeding one displacing the preceding, with essentially no interruption in flow. The acidic foam, for instance, can be displaced by foamed water into a vacuum truck containing a Dow Corning® silicone[8]* to collapse the foam. To prevent flash rusting, Brenner and Engle[10] suggest displacing the hydrochloric acid with a foamed 0.5 percent solution of citric acid (pH ~ 3), rather than with water. This can be followed by a combination neutralizing-passivating foam prepared from a solution containing 3 percent of Na_2CO_3, 1 percent of $NaNO_2$, 1 percent sodium citrate, and 0.5 percent of surfactant. After circulating for about an hour, the passivating foam may be collapsed in the vessel, then drained.

IX.4.d Vapor Phase Degreasing

In principle, vapor phase degreasing is the most appropriate method for removing organic fouling from columns, especially those used in petroleum processing. The procedure and a number of the solvents that can be used are discussed in Chapter III , pp. 88-92. Referring to Fig. IX.2, the overhead line from a small steam-heated still is connected by rigid piping to B and a suitable solvent, preferably one having a low enough boiling point to be vaporized easily, is introduced as a vapor into the top of the column to be cleaned. Here it condenses and flows downward through the trays, dissolving organic material, and collecting in the lower section of the column. The solution of organic material is then pumped from A into the still, where it is rectified, the vaporized solvent being recycled to the top of the column, and the organic fouling being concentrated in the still bottoms. When using flammable solvents all vessels must, of course, be grounded.

Eventually, all of the organic fouling will be transferred to the still, and

* Manufactured by Dow Corning Corporation, 2030 Dow Center, Midland, Michigan 48640

the reclaimed solvent can be recovered from the clean column. The organic residue can be washed out of the still with naphtha or mixed xylenes, and incinerated or reprocessed. Unless reclaimed, most solvents are too expensive to purchase and dispose of properly; this is especially true of 1,1,1-trichloroethane and o-dichlorobenzene. The principal advantage of vapor phase cleaning is that a large vessel can be efficiently cleaned with a relatively small volume of reclaimable solvent. A column having a capacity of 150,000 gal, for instance, can be cleared of organic fouling in 2-3 h by recirculating about 1000 gal of solvent, as outlined above.

IX.5 TYPICAL CHEMICAL CLEANING PROCEDURE

After the preceding generalities and summaries of available options, a specific procedure can now be developed for cleaning a contactor column used for removing H_2S from sour gas by absorption in monoethanolamine (MEA). This reaction, the neutralization of a weak acid (H_2S) by a weak base ($HOCH_2CH_2NH_2$), is readily reversed by increasing the temperature.[16]

$$HOCH_2CH_2NH_2 + H_2S \underset{240^\circ F}{\overset{100^\circ F}{\rightleftarrows}} HOCH_2CH_2NH_3HS \qquad \text{(IX-11)}$$

In normal operation, a dilute solution of MEA is pumped to the top of the contractor column, then flows downward through trays countercurrent to the sour gas, which is admitted at the bottom of the column. The sweetened gas is removed overhead while the rich MEA solution is heated and transferred to a regenerator where Eq. (IX-11) is reversed. As H_2S and other acidic gases are highly corrosive, relatively large amounts of FeS are likely to be found in contactors as the result of the direct reaction of H_2S with steel.

$$H_2S + Fe = FeS + H_2 \qquad \text{(IX-12)}$$

The shells of gas contactors usually are carbon steel, but the trays, which may be either sieve or bubble cap, are often stainless steel, in which case dilute inhibited sulfuric acid is the solution of choice for dissolving FeS. As the interiors of these vessels are nearly always fouled with oil, however, it is advisable to deoil before attempting to acid clean. Indeed, experience has shown that the deoiling process removes much of any FeS present as particulate matter, in some instances so effectively that further cleaning steps are unnecessary. This is because FeS formed as a corrosion product on a steel surface is not at all adherent and so much of it is dislodged and flushed out in the deoiling step.

The carbon steel column to be cleaned is 86 ft high, has an inside diameter of 7 ft, and contains 30 stainless steel bubble-cap trays. The capacity of the column, neglecting the volume of internal fittings is,

$$\pi(7)^2 \times 86/4 \cong 3310 \text{ ft}^3$$

$$3310 \times 7.48 \cong 25{,}000 \text{ gal}$$

As it is always desirable to minimize volumes of cleaning solutions, the cascading procedure outlined in Section IX.4.b, above, is selected so that the working volume of solution can be limited to about 4000 gal. Fig. IX.3 illustrates the necessary connections. Block valves in the line from the MEA reclaimer (not shown) are closed and the purified gas line is blinded. A 3-in. hose or screwed pipe connection is installed in the line that in operation carries MEA to the top of the contactor column. Cleaning solution is pumped through this line to the top of the column, cascades downward through the trays, and collects in the lower part of the column, from which it is then recirculated back to the top. A temporary 2-in. line is installed on the purified gas line to deliver any H_2S liberated to caustic scrubbers, or to an incinerator, if permitted.

NOTE: Reference should be made to Chapter VIII, pp. 165-171 and 176-181, before any attempt is made to chemically clean a column that may contain sulfides.

IX.5.a Deoiling

Time is always saved if a unit is turned over to the cleaning contractor with water in it. This can be conveniently done in most instances, as it is advisable to test a vessel hydrostatically for leaks before it is chemically cleaned; such a test is essential if there is any possibility of H_2S being evolved. The column should be drained, or filled, until it contains 4000 gal of water—about 14 ft in the column. Circulation is then started, the water being pumped from the bottom of the column through the piping labeled OUTLET in Fig. IX.3 to the top of the column through the connection labeled INLET. The pumping rate should be at least 450 gpm. Circulation is continued while the water is being heated by 50-psi steam in an external heat exchanger to about 140°F, at which time 110 gal of a 50° Bé solution of sodium hydroxide and 4 gal of a nonionic surfactant are added. Heating is continued until the temperature of the circulating solution reaches 150-160°F, during which time it is instructive to make Millipore filter tests every 30 min, as described in Chapter XI. This test is a measure of the undissolved solids circulating in the cleaning solution. In most cleaning operations the residue on the filter reaches a more-or-less constant weight, indicated by three consecutive results that do not differ by more than 5 percent; when this occurs, circulation is discontinued and the column is drained.

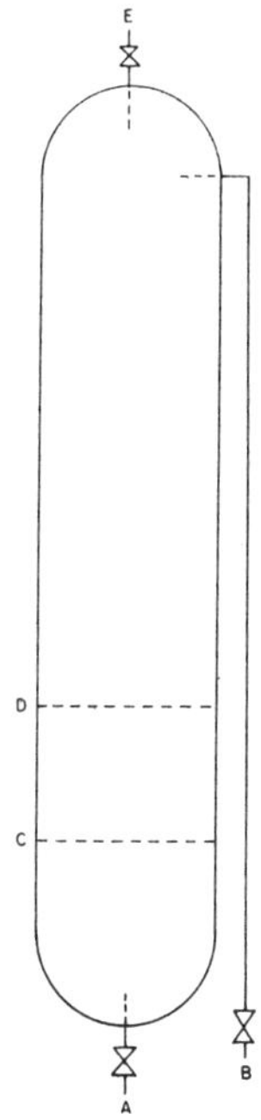

Fig. IX.3 Piping diagram for cleaning a column.

After about 30 min, representatives of the client and the contractor, wearing approved protective clothing, should inspect selected portions of the interior of the column to decide whether it is sufficiently clean, or also requires acid cleaning. In either event, 4000 gal of fresh water are added, and the cascading operation is continued for about 2 h, then the rinse water is drained. If no additional cleaning is deemed necessary, proceed to the passivation step described in Section IX.5.d, below.

IX.5.b Acid Cleaning

CAUTION: Before proceeding with acid cleaning make sure that adequate preparations have been made for the capture of any H_2S evolved, as described on pp. 165-171 and 176-181.

Using the blend-fill method described on p. 45, 4000 gal of inhibited 5 percent sulfuric acid are prepared. As indicated in Table II-5, p. 49, 120 gal of 66° Bé (93.2%) H_2SO_4 are required to prepare 4000 gal of 5 percent H_2SO_4. The solution is heated to 130-140°F, and cascaded through the column for 4-6 h at 200 gpm. The acid strength is checked every 30 min, and concentrated H_2SO_4 is added, as needed, to maintain the strength of the circulating solution at 5 percent. It may also be worthwhile to measure the concentration of total iron now and then to check the progress of the cleaning; Millipore filter tests also provide data that may prove useful in future cleanings.

After the cascading operation is completed, the acidic solution is drained into a vacuum truck, observing the precautions discussed in Chapter VIII. The column is next flushed, top to bottom, with fresh water until the pH of the rinse water is above 3.

IX.5.c Neutralization

Soda ash, Na_2CO_3, is added to raise the pH to about 8, and the solution is circulated without heating for 2 h; the pH is checked a few times during the neutralization to ensure that it does not fall much below 8.

IX.5.d Passivation

The carbon steel shell of the column under discussion is best passivated by flooding it with the passivating solution suggested by Duncan, et al.[17] This solution contains 0.25 percent each of Na_2HPO_4, NaH_2PO_4, and $NaNO_2$, and is prepared by dissolving 520 lb of each salt in the supply tank on the chemical cleaning truck. The salts can be added directly to the neutralizing solution or rinse water in the column, which is then filled with fresh water, heated to 200°F, and circulated for 15 min. After allowing the column to soak for an additional 30 min, the solution is drained and, after rinsing briefly, the column can be returned to service.

NOTE: Make sure that the contents of the column are neutral before admitting a solution containing sodium nitrite.

References

(1) McCoy, J. W. 1969. *The chemical anlaysis of industrial water.* New York: Chemical Publishing Co.

(2) Hollingsworth, R. L., Rowe, V. K., Oyen, F., Torkelson, T. R., and Adams, E. M. 1958. Toxicity of o-dichlorobenzene. *AMA Arch. Ind. Health* 17: 180.

(3) Sax, N. I. 1979. *Dangerous properties of industrial materials.* 5th ed. New York: Van Nostrand Reinhold Publishing Co.

(4) Plunkett, E. R. 1976. *Handbook of industrial toxicology.* New York: Chemical Publishing Co.

(5) Lawson, M. B., Martin, L. D., and Arnold, D. G. 1980. Chemical cleaning of FeS scales. Paper No. 219. *Corrosion/80* NACE. Chicago.

(6) Bell, W. E. 1965. Chelation chemistry–its importance to water treating and chemical cleaning. *Mater. Protect.* 4 (2): 78.

(7) Kowalski, X. 1980. *The DEQUEST® Phosphonates.* St. Louis: Monsanto Industrial Chemicals Co.

(8) Mayer, L. and Guelich, J. Oct., 1963. Hydrogen fluoride (HF) inhalation and burns. *Arch. Environ. Health* 7 (4): 71.

(9) Jones, C. C. 1973. Nickel carbonyl poisoning. *Arch. Environ. Health* 26: 245.

(10) Brenner, J. W. and Engle, J. P. June 5, 1962. *Foam cleaning of surfaces.* U.S. Patent No. 3,037,887.

(11) Carroll, D. B. and Rodgers, J. W. Oct. 19, 1965. *Foam generator.* U.S. Patent No. 3,212,762.

(12) Crowe, C. W. and Duranleau, R. G. Apr. 1, 1969. *Cleaning by foam contact and foam regeneration method.* U.S. Patent No. 3,436,262.

(13) Simmons, B. G. Jan. 9, 1979. *Apparatus for making foamed cleaning solutions and method of operation.* U.S. Patent No. 4,133,773.

(14) Zimmerman, O.T. and Lavine, I. 1960. *Handbook of material trade names.* Supplement III. Dover, New Hampshire: Industrial Research Service Inc.

(15) Hackh's chemical dictionary. 1969. J. Grant, ed. New York: McGraw-Hill Book Co.

(16) *Gas conditioning fact book.* 1962. Midland, Michigan: Dow Chemical Co.

(17) Duncan, F. R., Fyock, D. H., Baldwin, L. V., and Hopkins, R. D. 1972. Preoperational cleaning of high pressure boilers of General Public Utilities Corporation. *Proc. Int'l. Water Conf., Eng. Soc. W. Pa.* 33: 97.

X

Furnace Tubes

For purposes of this brief discussion industrial furnaces are classified into two groups: (1) preheat furnaces in which petroleum or other material is heated for some purpose such as distillation or to increase the rate of a chemical reaction; (2) integral-furnace boilers consisting of a firebox for heating water in a steam generator, both furnace and steam generator being combined in a single structure. In general, fouling in a preheat furnace occurs within the tubes, whereas in an integral-furnace boiler, slags and soot coat the outsides of the tubes. Instances are encountered, of course, where the reverse is true, for example, water-formed deposits in boiler tubes; this aspect is covered in Chapter VII.

X.I PREHEAT FURNACES

A gas- or fuel oil-fired furnace is used to heat crude petroleum to 650–700°F preparatory to atmospheric distillation. This is accomplished by passing the oil through several rows of tubes 4–6 in. in diameter enclosed in a large firebox heated by natural gas or fuel oil burners located on the bottom or sides of the furnace. In a five-pass furnace, for instance, there are five parallel flow paths, which serve to reduce the pressure drop associated with two-phase flow when partial vaporization occurs in the tubes. The convection section is in the upper part and the radiant section is in the lower part of these furnaces. With the exception of one or two preheat exchanges, essentially all of the heat to an atmospheric column is supplied by the furnace so care must be taken to limit the temperature of the feed to the column to 700°F in order to avoid coking associated with thermal cracking, which commences at about 750°F. As it is impossible to attain 100 percent salt removal in a commercial desalter, residual salt tends to deposit on the furnace tubes, reducing the rate of heat transfer and necessitating higher firing rates to maintain throughput; this in turn lays down coke. Other common foulants that must be removed from time to time are FeS, FeS_2, and Fe_2O_3.

X.1.a Steam-Air Decoking

Coke formed in a furnace tube over a relatively long period of time is dense and hard, but if quickly formed it tends to be porous. The latter type is easily removed by dry firing the furnace at about 700°F while passing 150-psi steam through the tubes with air being added at such a rate that the concentration of oxygen is about 1 percent in the steam. The course of the coke removal can be followed by observing the progress of a red-hot ring along the tubes; this can be seen from outside the furnace. Coke removal is more nearly complete if the air supply is interrupted occasionally and the flow of steam is increased to take advantage of the abrasive effect of any spalled coke. Porous coke up to 2 in. thick is readily oxidized by steam-air decoking, but a thin layer of shell coke often remains that is extremely difficult to remove. Exit gases are quenched in cold water, which first becomes black with particles of carbon, later gray, and finally bright red with particles of hematite, Fe_2O_3. This is the most economical procedure for removing coke from preheater furnace tubes and also has the advantage of rendering FeS_2 soluble in acids by oxidizing it to Fe_2O_3 and SO_2.

$$2FeS_2 + 7/2O_2 = Fe_2O_3 + 2SO_2 \qquad \text{(X-1)}$$

X.1.b Nitrogen Blasting

Porous coke can be jetted from furnace tubes using a line mole with water at about 6000 psi, but to remove shell coke the more rigorous procedure described by Nunciato, et al.,[1] in which steel shot is entrained in nitrogen moving at high velocity, must be used. Steel shot is less abrasive to tube metal than sand or flint, but if the tubes are at all oily, the shot sticks to them and cleaning is poor. In this case, the furnace should be fired for a while with a few burners, say four on each side, to dry the tubes before shot blasting. The success of cleaning can be measured by determining furnace efficiency before and after cleaning; monitoring pressure drop through the tubes or x-raying them are also useful measures.

X.1.c Chemical Cleaning

The mixture of FeS, FeS_2, Fe_2O_3, and coke laid down in preheat furnaces is virtually insoluble in any solvent but, as noted in Section X.1.a, it becomes acid-soluble in the steam-air decoking process. Although substantial amounts of the inorganic portion are loosened and flushed out by the steam, it may be necessary to follow with an acid cleaning. If this is done, it is necessary to make provision for the possible evolution of H_2S, as described on pp. 165-171 and 171-181 , even though most of any sulfides present will have been oxidized to SO_2 during the steam-air decoking. Chemical cleaning is carried out much as described in previous chapters, i.e., a suitable inhibited acid is circulated through the tubes, after which they are rinsed, neutralized, and passivated. Note that if preheat furnace tubes are acidified without previously steam-air decoking, copious quantities of H_2S are evolved. In this application alkali scrubbers must be used as well as rigid piping, preferably seal welded.

X.2 INTEGRAL-FURNACE BOILERS

Furnaces in which fuel oil is burned are subject to corrosion and the formation of fused slags arising from the presence of certain chemical elements, notably nickel and vanadium, in the fuel oil. As these difficulties are not nearly so severe when coal is used as fuel, and are essentially nonexistent with natural gas, the cleaning of the fire side of furnace tubes is confined, in the main, to removing slags associated with burning fuel oil.

Beach and Shewmaker,[2] as a result of extraction studies, have established two classes of vanadium compounds in petroleum. Class I contains those that are extractable with aqueous pyridine; Class II contains those that are not extractable. By conducting molecular distillations using castor oil as a carrier, they estimated the equivalent atmospheric boiling points of the volatile porphyrins to be in the range of 1085-1200°F, with molecular weights of 543-800. They conclude from measurements of absorbancies that the difference in the volatile and nonvolatile aggregates is probably in the peripheral groups rather than in the nitrogen-vanadium bonds, and that the nonvolatile porphyrins have large asphaltic or polymeric side chains that distort the nitrogen-vanadium bond, thus producing a noncharacteristic spectrum. Although vanadium is associated particularly with asphalt or asphalt-containing crudes, practically all crude oils contain some vanadium, its concentration ranging from a few tenths to several hundred parts per million. The element tends to concentrate in fuel oils and is converted to various slag-forming compounds in the process of burning.

Many different compounds have been reported in furnace slags, the most prevalent of which are listed in Table I-4 , p. 13. The application of energy-dispersive x-ray fluorescence and refined methods of x-ray diffraction has shown the presence of a number of other more complex compounds, among them $(NH_4)_3(Fe,V)(SO_4)_3$, $(Ni,Fe)_3V_2O_8$, $(Fe,Ni,V)_2O_3$, and $(Fe,Ni,V)_3O_4$.* See Chapter I , pp. 8–15 for additional information on the mechanism of slag formation.

Salooja[3] has reviewed the additives used to inhibit the formation of slag, soot, and smoke emission, high-temperature corrosion by fused salts, and low-temperature corrosion by sulfuric acid. For many years inorganic oxides have been added to fuel oils to raise the melting points of vanadium salts and to neutralize acidity produced by sulfur. Magnesium oxide, hydroxide, carbonate, and sulfate, and dolomitic lime are the materials that have been most widely used to induce the formation of loose powdery deposits instead of slags. These additives also convert sulfur to stable sulfates, thus reducing corrosion in cooler sections of furnaces. Dolomite lime and solid inorganic compounds are applied as a slurry prepared by mixing 100 lb of finely ground chemical with 180 gal of heated light fuel oil, which is then pumped continuously into the hot fuel oil line to the burners. These additives have the disadvantage of raising the ash content of the fuel oil, increasing particulate emissions from the stack, and eroding burner tips.

In an elaborate investigation, Lee et al.[4] found the most effective and economical ash modifying mixture to be 10.7 percent of MgO and 10.7 percent of Al_2O_3, with particle sizes in the range of 2–7 microns, suspended in a light oil and stabilized with 0.5–2.0 percent of a surfactant. This combination formed a porous, friable, flaky deposit with a cubic structure softening above 2800°F. They found the most effective dosage to be 1.75 parts of ($MgO + Al_2O_3$) to 1.0 part of vanadium, and suggest that the reaction between additive and fuel oil ash probably occurs at the tube surface rather than in the vapor phase. Ash formed this way consisted of layers of unreacted additive, additive–vanadium reaction products, and sodium vanadyl vanadate, $Na_2O \cdot V_2O_4 \cdot 5V_2O_5$, dispersed in the additive oxides. Other ash modifiers that have been proposed are zinc oxide, manganous oxide, and calcium oxide.

* O. L. Davis, Chevron Research Co., private communication.

X.2.a Hydraulic Cleaning

Cleaning of the outside (fire side) of water tubes in integral-furnace boilers is always done either mechanically or by high-pressure water jetting. Various specialized types of equipment have been designed for the latter application including rotating heads for water jetting relatively inaccessible places and clinker jets for cleaning coal-fired furnaces. In use, the clinker jet forces water at about 6000 psi into hot clinkers where steam is formed, shattering the encrustation. This operation is sometimes done on-stream in utility boilers, as is sand blasting, but both create more particulate emissions than are acceptable to air pollution authorities.

X.2.b Removal of Slag and Fused Salts

In most instances, furnaces are cleared of slag by mechanical chipping. Lee, et al.,[4] however, found that furnace slag is bound to the superheater and other tubes by water-soluble sulfates. Slag itself contains a great number of small interconnecting pores, which although impermeable to water alone, are permeated by water containing 0.1-0.2 percent of a nonionic wetting agent. By saturating the slag in a furnace with water containing a wetting agent, and allowing it to stand for a day or two, part of the binding salt is dissolved, and the bulk of the slag becomes susceptible to removal by high-pressure water jetting. This simple procedure, where applicable, saves considerable labor.

References

(1) Nunciato, D. J., White, N. H., and Woodburn, W. A. May 20, 1980. *Method for decoking fired heater tubes.* U.S. Patent No. 4,203,778.

(2) Beach, L. K. and Shewmaker, J. E. 1957. The nature of vanadium in petroleum. *Ind. Eng. Chem.* 49: 1157.

(3) Salooja, K. C. Jan., 1973. Burner fuel additives. *Combustion* 44 (7): 21.

(4) Lee, G. K., Mitchell, E. R., Grimsey, R. G., and Hopkins, S. E. 1964. An investigation of fuel-oil additives to prevent superheater slagging in naval boilers. *Proc. Am. Power Conf.* 26: 531.

XI

Analytical Procedures

Chemical analysis plays, or at any rate should play, an important role in chemical cleaning operations. The strengths of cleaning solvents must be continually determined during the process of blend-filling a vessel and throughout the course of a chemical cleaning procedure. The single most important criterion for selecting a solvent is the solubility of a fouling deposit in that solvent, determined with a sample that is, to some extent at least, representative of the bulk of the material to be removed. A complete chemical analysis is seldom needed, but it is useful to have an estimate of the percentage of organic and inorganic material in a deposit so that an informed judgment can be made as to the advisability of including a preliminary deoiling step. Methods for chemically analysing scales, sludges, and deposits are available in another publication.[1] Emission spectroscopy is often used for analyzing fouling deposits, although it is seldom possible to make quantitative determinations of the principal metallic components more precisely than ± 10–15 percent. In general, components that are present in concentrations less than 1 percent are more accurately determined spectrographically, but these ordinarily are of slight interest in selecting a chemical cleaning procedure. X-ray diffraction has been applied with some success to water-formed deposits for the identification of chemical compounds.[2] When many different crystal forms are present, however, the analysis of the diffraction patterns is extremely complex, and for this reason the method is not especially applicable to corrosion deposits or to those containing a variety of elements. A standard method has been developed for analyzing water-formed deposits by x-ray fluorescence.[3]

For purposes of chemical cleaning rapid methods of analysis that give reasonably reliable results are in order; using this approach, results are obtained at very little cost that are adequate for both the chemical cleaning contractor and the coordinator. In describing various procedures the word determination is used rather loosely throughout this chapter, and it should be understood that in most instances estimation would be a more appropriate term.

XI.1 SAMPLING AND CHARACTERIZATION OF DEPOSITS

In Chapter I, pp. 1-15, various types of deposits and methods of sampling them are discussed in a general way. In this and the next section directions are given for characterizing a deposit and finding a suitable solvent for removing it. The original sample is most often collected by a representative of an operating plant, preferably in consultation with the chemical cleaning coordinator. The sample is then sent to a laboratory, either that of the chemical cleaning contractor or the client's own facility, for some sort of analysis.

Scales are sampled by scraping with a sharp steel blade, taking care not to introduce any metallic chips from the blade or substrate. Multilayered scales, such as occur in furnaces, may be segregated according to layers at the site of sampling, if desired. If it is possible to dislodge a fairly large piece, however, it is more convenient to separate the layers in the laboratory.

Wet sludges can be separated from the bulk of their water by straining through cloth, dipping with a sieve, or collecting on paper towels. Wet samples and those from corrosive environments should not be stored in cans for the container is rapidly attacked. For the majority of samples a wide-mouth Pyrex bottle is a suitable container.

The sample as received in the laboratory should represent, so far as is possible, the material to be removed by chemical cleaning. If the sample is partly organic and partly inorganic it is helpful to know the proportion of each when selecting the method of cleaning. Accordingly, samples should be dried briefly, then extracted with a suitable solvent to measure the approximate proportions of organic and inorganic materials. Samples should not be submitted containing large amounts of water, acids, alkalis, petroleum products, or other extraneous material not considered part of the deposit, without advising the analyst. Any oil in a sample obtained from a boiler or cooling system should, of course, be extracted and weighed after preliminary drying, as oil is obviously foreign to these systems, and provision must be made for removing it in the chemical cleaning procedure.

Deposits from furnaces seldom contain a significant amount of water, so it is usually unnecessary to dry the sample. Furnace slags consisting of several laminae must be separated by hand if it is desired to determine the effect of solvents on each layer. Scales and slags from the radiant and superheater sections of furnaces are ordinarily free of carbon and may be treated directly by the procedure in Section XI.2. Samples that do contain carbon should always be extracted with benzene, however, as free carbon adsorbs organic material, including unburned fuel oil and heavy sulfonic acids that increase the difficulty of removal by chemical cleaning.

Accumulations from boiler drums are generally in the form of mud or sludge containing more or less water, which should be decanted and discarded. Boiler deposits sometimes contain oil introduced in contaminated condensate. When oil is present in boiler water loose scale may form, particularly in water-wall tubes and headers, the oil serving as a nucleus and binder for scaling at hot spots. As most scales from boiler tubes are dry and extremely insoluble, the drying and extraction with benzene can often be omitted. Another source of organic fouling in boilers is the attrition of resins used for ion exchange; these resin particles do not dissolve in many solvents, but they can easily be washed away after the bulk of associated fouling has been disintegrated.

Rarely is it possible to obtain samples of fouling for anlaysis at the time a chemical cleaning becomes necessary. During scheduled shut downs, however, samples should be collected for analysis and solubility characterization, documenting the source, composition, and solubility of each sample. Table XI-1 lists the most common components of fouling deposits in various types of industrial equipment. For similar listings see Table I-1, p. 5, Tables I-4 and I-5, pp. 13–14, and Table VI-1, p. 117.

The following procedure provides general directions for drying and solvent-extracting the laboratory sample; to accomplish this a weighed portion of the sample is oven-dried and then extracted with benzene in a Soxhlet extractor. Benzene is chosen because it readily dissolves most organic material likely to be encountered, including oil, oxidized oil, greases, tars, and asphalts, as well as free sulfur, leaving behind essentially all inorganic material. Deposits containing oxidized oil often also contain very finely divided carbon that passes through the pores of the extraction

TABLE XI-1

Components of Fouling Material in Industrial Equipment

Equipment or Vessel	*Common Components*
Boilers (water side)	Fe_2O_3, Fe_3O_4, Al_2O_3, SiO_2, $MgSiO_3$, $CaSO_4$, CuO, $Ca_3(PO_4)_2$, oil, organics
Heat exchangers (water side)	Fe_2O_3, $FePO_4$, $CaCO_3$, $Ca_3(PO_4)_2$, $MgSiO_3$, Al_2O_3, SiO_2, Cr_2O_3, CuO, ZnO, oil, organics, biological
Surface condensers (water side)	Fe_2O_3, Fe_3O_4, $CaCO_3$, $CaSO_4$, oil, microbiological
Compressor jackets (water side)	Fe_2O_3, Fe_3O_4, $CaCO_3$, $Ca_3(PO_4)_2$, oil, biological
Heat exchangers (process side)	FeS, Fe_2O_3, Fe_3O_4, oil, organics
Furnace tubes (process side)	FeS, FeS_2, Fe_2O_3, Fe_3O_4, coke, oil, organics
Furnace tubes (fire side)	Na_2SO_4, $NiSO_4$, $2NiO \cdot V_2O_5$, $Na_4V_2O_7$, Na_3VO_4, NiO
Fractionating columns	FeS, Fe_2O_3, Fe_3O_4, oil, asphalts, tars, organics
Reboilers	Fe_2O_3, Fe_3O_4, FeS, FeS_2, oil, organics

thimble. The amount lost is small, and no provision is made for recovering it. For special purposes, for instance, to identify the source of oil contamination in a cooling system, it may be desirable to submit the benzene extract to chromatographic analysis.[4]

XI.1.a Characterization of Fouling Deposits

TEST

Procedure: Remove any excess water or oil from the original material by filtration or decantation, and discard it.*

Mix the residue thoroughly with a rod or spatula, weigh a representative

* In addition to water and oil, other extraneous matter such as pieces of wood, glass, etc., or excess process chemicals may be removed at the discretion of the analyst.

10-g portion, and transfer it to a Soxhlet extraction thimble of suitable size. Place the thimble upright in a beaker, and dry in an oven at 105°C for 1 h. Cool and reweigh, noting the loss in weight.*

Place the thimble in an extractor, add 200 ml of benzene to the flask, and reflux for 1 h, or until the refluxing benzene is colorless.

Cool the apparatus, remove the thimble, and place it in a beaker on a steam-plate until the odor of benzene is no longer detectable. Dry briefly in the oven and reweigh the thimble and extracted sample.**

From the losses in weight calculate and report percent water; percent benzene-soluble (organic); percent inorganic material.

XI.2 ESTIMATION OF SOLUBILITIES

The purpose of the extraction procedure in Section XI.1 is to characterize a fouling deposit in such a way that by applying a few elementary principles, a suitable solvent can be selected for removing the material, without performing a large number of solubility tests. The process is best illustrated by a specific example.

Suppose, through misadventure, that an 850-psi boiler associated with a catalytic cracking plant becomes contaminated with some kind of oil, which upon opening the boiler for inspection, is found to be intimately mixed with granular material, forming a sticky combination that is distributed throughout the mud drum and lower headers. Upon subjecting the material to the procedure in Section XI.1, the following results are obtained:

Water	1.4 percent
Benzene-soluble	63.5 percent
Inorganic residue	35.1 percent

A chromatographic analysis of the benzene extract suggests that the organic portion of the fouling is primarily medium cycle oil, an intermediate product in catalytic cracking having a boiling range of 460–720°F, corresponding to carbon numbers in the range of C_{12}-C_{26}. As aromatic compounds are resistant to catalytic cracking, they tend to accumulate in the reactor product, a portion of which is recycled in an effort to obtain more complete conversion and improve yield. These aromatic mixtures are, for this

* Samples containing light hydrocarbons or solvents should first be placed on a steam-plate for a few minutes to minimize the danger of an explosion in an oven with open heating elements.

** The loss in weight should be reported as "benzene-soluble."

reason, called cycle oils.

As a rule, it is unnecessary to obtain a complete chemical analysis of the inorganic residue, but for purposes of this illustration it was done, with results shown in Table XI-2.

TABLE XI-2

Chemical Analysis of Inorganic Residue

Component	*Percent by weight*
CaO	1.0
Cr_2O_3	0.8
CuO	23.7
Fe_3O_4	72.2
MgO	0.2
NiO	0.2
PO_4	1.4
ZnO	0.4

Having ascertained that the organic portion of the boiler fouling is medium cycle oil, several solvents can be eliminated from consideration. As medium cycle oil contains a high proportion of high-boiling aromatics, sodium hydroxide alone would have little effect as the material is not susceptible to saponification, nor is it easily emulsified. Similarly, salts of weak acids such as phosphate and silicate produce much lower concentrations of hydroxyl ion than does sodium hydroxide, so these also would be ineffective as solvents. Because of the high proportion of heavy oil in the deposit, chelants such as EDTA or NTA would not be appropriate solvents; for the same reason, neither an alkaline oxidant nor an acid alone would be likely to be successful as cleaning solvents. Sodium hydroxide combined with a so-called inorganic builder, e.g., sodium phosphate or sodium meta-silicate usually performs better with organic fouling than do any of these compounds alone. When using phosphate with sodium hydroxide it should be noted that there is no point in using any phosphate except Na_3PO_4, as the dibasic and monobasic salts react with excess NaOH to form the tribasic salt.

$$HPO_4^{--} + OH^- = PO_4^{-3} + H_2O \qquad \text{(XI-1)}$$

$$H_2PO_4^- + 2OH^- = PO_4^{-3} + 2H_2O \qquad \text{(XI-2)}$$

An alkaline emulsion of NaOH and automotive diesel fuel might perform reasonably well, but one prepared with mixed xylenes should be considerably better in this case, as the organic material to be dissolved is also highly aromatic. Acidic emulsions should also be fairly efficient. Mixed xylenes alone would dissolve essentially all of the organic material, but would have no effect on the inorganic components. Finally, a two-step procedure in which the system is first deoiled, then treated with acid, almost always yields the best results. Table XI-3 is a summary of the actual solubilities obtained with a variety of solvents applied as described in the procedure below.

TABLE XI-3

Solubilities of a Fouling Deposit in Various Solvents

Test	*Formulation*	*Type of solvent*	*Percent soluble*
1	5% NaOH + 0.2% surfactant	Alkaline	9
2	0.5% Na_3PO_4 + 0.1% Na_2HPO_4 + 0.2% surfactant	Alkaline	15
3	2% EDTA @ pH 3.5	Chelant	17
4	3% NaOH + 2% $KMnO_4$	Alkaline oxidant	24
5	10% H_2SO_4 + 0.2% surfactant	Acidic	26
6	2% NaOH + 0.5% Na_3PO_4	Alkaline	33
7	20% Auto diesel + 80% of 5% NaOH	Alkaline emulsion	35
8	50% Auto diesel + 50% of 5% NaOH	Alkaline emulsion	37
9	20% Auto diesel + 80% of 15% HCl	Acidic emulsion	57
10	80% Mixed xylenes + 20% of 5% NaOH	Alkaline emulsion	65
11	100% Mixed xylenes	Aromatic hydrocarbon	65
12	50% Auto diesel + 50% of 15% HCl	Acidic emulsion	70
13	80% Mixed xylenes + 20% of 10% H_2SO_4	Acidic emulsion	75
14	Two-stage: a) Mixed xylenes; b) 15% HCl	Aromatic; acidic	96

Note: All solubility tests were carried out at 160–175°F for 6 h.

It can be said, as a generality, that a deposit containing more than 30 percent of organic material is not likely to be successfully removed by chemical cleaning unless an organic solvent is included in the formulation. If 40-50 percent of an inorganic deposit can be dissolved in some suitable solvent, chemical cleaning will probably be successful. In some instances, even though the total solubility is small, it may be that binders holding insoluble material together in a hard scale are selectively dissolved, making the material transportable, e.g., by flushing or water jetting. Experience has shown that results in the field are often better than predicted by laboratory analyses. In selecting a solvent metallurgical considerations must not be overlooked: obviously, there is no point in determining solubility in a solvent that is incompatible with any of the materials of construction of the unit to be cleaned. Also, in this particular example, the presence of a substantial amount of copper requires the application of one of the special procedures described in Chapter VII, pp. 150-153, after deoiling.

XI.2.a Estimation of Solubility

ESTIMATION OF SOLUBILITY TEST

Equipment: *Condenser Tube.* ₮ 24/40, 650 mm. Catalog No. 52300-650. Available from Science Products Division, Corning Glass Works, Corning, New York 14830.
Erlenmeyer Flask. Narrow mouth, ₮24/40, 125-ml. Corning Catalog No. 5000-125.
Gooch Crucible. 30-ml, with coarse fritted disc. Corning Catalog No. 32940-30C.

Procedure: Accurately weigh approximately 1 g of fouling material, as received, and transfer it to a 125-ml Erlenmeyer flask.

Add 100 ml of solvent to be tested, and swirl the contents briefly, observing the effect of the solvent on the fouling material.*

Attach an air condenser to the flask, then transfer the assembly to a steam-plate or a thermostatically controlled hot-plate, and heat for 4-6 h at 160-175°F.

Remove the assembly from the heat, allow to cool, then pour the solution through a tared Gooch crucible, transferring any undissolved residue to the crucible with a suitable wash solution.

---

* Note whether H_2S is evolved, sample tends to disintegrate, etc.

Dry the crucible in an oven at 105°C for 30 min, cool, and reweigh.

$$\% \text{ soluble} = \frac{\langle \text{sample weight} - [(\text{crucible} + \text{residue}) - \text{tare}] \rangle \times 100}{\text{sample weight}}$$

XI.3 SPECIAL FIELD PROCEDURES

Included in this section are three useful procedures that are occasionally needed and that can be performed with simple equipment at the job site. The first of these is a quick and reliable test that shows whether corrosion inhibitor has been added to an acid cleaning solution. It is sometimes desirable to measure, by a Millipore filtration, the amount of solid material removed during a deoiling procedure applied to a vessel that has been found in previous cleaning to contain nonadherent deposits. An example of this is discussed on p. 207; the method for performing the filtration is described below. During the course of an acid cleaning in which H_2S is being evolved and captured in a caustic scrubber, it is essential that free alkali be always available to neutralize the acid gas. A procedure for checking the concentration of free alkali in the presence of carbonate and sulfide is also described in this section.

XI.3.a Steel Wool Test for Inhibited Acid

Filming amines are corrosion inhibitors that prevent acidic cleaning solutions from attacking metal without interfering with the dissolution of scales and deposits. They form a chemisorbed monomolecular film on the surface of metal that impedes the reduction of hydrogen ions and the oxidation of metal. Filming amines are, thus, general corrosion inhibitors that stifle both the anodic and cathodic half-reactions. As steel rapidly dissolves in hot, uninhibited mineral acids, the chemical cleaning contractor must ensure that inhibitor has been added according to the manufacturer's instructions before introducing an acidic cleaning solvent into industrial equipment. Although some commercial inhibitors contain blue or red dyes that are obvious when added to acidic solutions, others are not so tagged. The following procedure provides a simple and convincing demonstration of the presence of corrosion inhibitor when there is doubt whether an acidic cleaning solution has been inhibited:

CORROSION INHIBITOR TEST

Reagent: *Steel Wool.* Washed with petroleum ether to remove oil, and dried briefly in air.

Procedure: Place a small wad of steel wool in each of two 150-ml beakers.

Transfer a 100-ml portion of the acid in question to one of the beakers, then add a 100-ml portion of uninhibited acid of the same strength to the other.

Bubbles of hydrogen will immediately begin to evolve within the wad of steel wool in the uninhibited acid, and within a few seconds it will rise to the surface.

If the acid in question is inhibited, the wad of steel wool will remain on the bottom of the beaker, as no hydrogen will be evolved.

XI.3.b Millipore Filter Test for Insolubles

In order to measure the progress of the deoiling step in a chemical cleaning the weight of insoluble material circulating in the deoiling solution can be determined by filtering a measured portion of the solution through a Millipore filter, and weighing the residue. The sample is taken from a sample draw in the return line to the contractor's truck every 30 min or so, until the weight of residue does not increase by more than 5 percent in each of three successive samples, at which time deoiling is judged to be complete. If, for example, successive residues were obtained weighing 0.235, 0.247, 0.259, and 0.272 g, deoiling would be considered complete and the contractor would proceed to the next phase. If these residues were obtained from 250-ml portions of deoiling solvent, as specified in the following procedure, and the total volume of the solution was 20,000 gal, the total weight of insoluble material circulating can be easily calculated. Taking the average of the last three weights as 0.259 g, the total weight of insoluble material dislodged is:

$$\frac{0.259\ (\text{g}/250\ \text{ml}) \times 1000\ (\text{ml/l}) \times 3.8\ (\text{l/gal}) \times 20{,}000\ (\text{gal})}{454\ (\text{g/lb})} \cong 173\ \text{lb}$$

MILLIPORE FILTER TEST

Equipment: *Millipore All-Glass Filter Apparatus,* 47 mm. Catalog No. XX 15 047 00, Millipore Corporation, Bedford, Massachusetts 01730.
Fluoropore Filters, 1.0 μm. Catalog No. FALP 047 00.

Vacuum Pump. A water aspirator or other source of vacuum can also be used.

Procedure: Collect a 250-ml sample of the recirculating deoiling solution, and transfer it to the funnel of the Millipore Filter Apparatus that has been fitted with a 47-mm, 1.0 μm Fluoropore filter previously weighed to the nearest milligram.

Transfer the residue to the filter using a wash bottle, rinse the residue on the filter with 2 or 3 10-ml portions of water, then aspirate the filter for 15 min to dry it.

Reweigh the filter and, if desired, calculate the pounds of insoluble material circulating in the solution, as in the preceding discussion.

XI.3.c Free Alkali in H_2S Scrubbers

The use of caustic scrubbers to neutralize H_2S evolved during chemical cleaning is described in Chapter VIII, pp. 165–171 and pp. 176–181. As explained on pp. 169-170, cleaning acid and sodium hydroxide are balanced so that at least 10 percent of sodium hydroxide remains in the secondary scrubber when all of the acid has been consumed. In the interests of safety and security of the workmen, however, it is advisable to measure the free NaOH in both the primary and secondary scrubbers from time to time during the acid cleaning step. In addition to sulfide and hydroxyl ions, alkaline scrubbers often contain carbonate ion arising from carbon dioxide evolved from calcium carbonate scales; on infrequent occasions mercaptans may also be found. The precise analysis of a solution containing all of these components, at best a complicated undertaking, has been discussed in another publication.[1] For monitoring the free sodium hydroxide only, a relatively simple procedure suffices.

In the course of an acid cleaning to remove sulfide deposits, hydrogen is evolved, and is captured by reaction with excess NaOH.

$$H_2S + 2OH^- = S^{--} + 2H_2O \qquad \text{(XI-3)}$$

As the supply of hydroxyl ion diminishes, hydrosulfide begins to form:

$$H_2S + OH^- = HS^- + H_2O \qquad \text{(XI-4)}$$

When most of the hydroxyl ion has been neutralized, sulfide also reacts with hydrogen sulfide:

$$H_2S + S^{--} = 2HS^- \qquad (XI\text{-}5)$$

Carbon dioxide liberated in the acidification of carbonates reacts analogously:

$$CO_2 + 2OH^- = CO_3^{--} + H_2O \qquad (XI\text{-}6)$$

$$CO_2 + OH^- = HCO_3^- \qquad (XI\text{-}7)$$

$$CO_2 + CO_3^{--} + H_2O = 2HCO_3^- \qquad (XI\text{-}8)$$

In addition to these neutralizations, sulfide and carbonate both hydrolyze:

$$S^{--} + H_2O = HS^- + OH^- \qquad (XI\text{-}9)$$

$$CO_3^{--} + H_2O = HCO_3^- + OH^- \qquad (XI\text{-}10)$$

In measuring free NaOH, the alkalinity produced by hydrolysis should not be included, so steps must be taken to reverse these two reactions before titrating the free NaOH remaining in the scrubbers. This can be done by precipitating carbonate and sulfide by adding barium chloride followed by zinc chloride solutions:

$$Ba^{++} + CO_3^{--} = BaCO_3 \qquad (XI\text{-}11)$$

After filtering the solution to remove barium carbonate the zinc solution is added to the filtrate:

$$Zn^{++} + S^{--} = ZnS \qquad (XI\text{-}12)$$

$$Zn^{++} + HS^- = ZnS + H^+ \qquad (XI\text{-}13)$$

By this means alkalinity formed in the hydrolysis of carbonate and sulfide disappears. It is apparent, too, that HS^-, if present, liberates hydrogen ion upon the addition of zinc ion. If the scrubber solution contains free NaOH, however, the concentration of HS^- will be negligible because the reaction represented by Eq. (XI-3) predominates. The successful titration of free alkalinity also requires the presence of ammonium ion to prevent the precipitation of hydroxyl ion by excess zinc salt:

$$Zn^{++} + 2OH^- = Zn(OH)_2 \qquad (XI\text{-}14)$$

$$Zn^{++} + 4NH_4OH = Zn(NH_3)_4^{++} + 4H_2O \qquad (XI\text{-}15)$$

In the presence of ammonium ion an acid-base indicator must be chosen that changes color between pH 4 and 5, the approximate acidity produced by the hydrolysis of ammonium salts:

$$NH_4^+ + H_2O = NH_4OH + H^+ \qquad \text{(XI-16)}$$

The hydrolysis constant for this reaction can be calculated from the dissociation constant of ammonium hydroxide and the ion product of water

$$K_h = \frac{(NH_4OH)(H^+)}{(NH_4^+)} = 1 \times 10^{-14}/1.7 \times 10^{-5} = 5.9 \times 10^{-10}$$

As the concentrations of ammonium hydroxide and hydrogen ion are equal, and the concentration of ammonium ion is very large compared to theirs, namely, about 2 moles per liter, the following can be written:

$$K_h = (H^+)^2/2 = 5.9 \times 10^{-10}$$

$$(H^+)^2 = 2 \times 5.9 \times 10^{-10}$$

$$(H^+) = 3.4 \times 10^{-5}$$

$$pH = 4.5$$

Bromocresol green changes from yellow to blue over the pH range of 4.0–5.4, and is not disturbed by sulfide ion as are some other indicators (methyl orange, alizarin red, bromothymol blue) that would otherwise be suitable.

In those rare instances when mercaptans are encountered, their precipitation by zinc ion is only about 75 percent complete. Precipitation can be made complete by adding cadmium chloride, but this salt cannot be used for precipitating sulfide, as the yellow CdS obscures the yellow end point obtained in the titration of free alkalinity.

It should be mentioned that free NaOH can be measured approximately by titrating a sample to thymolphthalein, which changes from colorless to blue in the pH range of 9.4–10.6, but the following procedure is much more reliable for partially spent scrubbing solutions.

FREE ALKALI IN CAUSTIC SCRUBBERS TEST

Reagents: *Barium Chloride,* 10%. Dissolve 100 g of $BaCl_2 \cdot 2H_2O$ in 1000 ml of water and neutralize to phenolphthalein with 0.1 *N* NaOH.

Bromocresol Green Indicator. Dissolve 0.4 g of 3′,3″,5′,5″-tetrabromo-m-

cresol-sulfonaphthalein (Eastman No. 1782) in 5.7 ml of 0.1 *N* NaOH, and dilute to 1000 ml.

Zinc Chloride Solution. Dissolve 32 g of $ZnCl_2$ and 105 g of NH_4Cl in about 500 ml of water. Add 65 ml of bromocresol green indicator solution, and dilute to 1000 ml.

Cadmium Chloride Solution. Dissolve 380 g of $CdCl_2 \cdot 2.5H_2O$ in about 600 ml of water. Add 25 ml of bromocresol green indicator solution and 2.5 ml of 0.1 *N* NaOH, and dilute to 1000 ml.

Procedure: By means of a small hydrometer, measure the specific gravity of the sample.*

Measure a 2.0-ml sample with a pipet and transfer it to a 50-ml glass-stoppered mixing cylinder. Make up to volume with 10 percent $BaCl_2$.

Shake the mixture vigorously, filter the solution through a Whatman No. 12 folded filter paper, and transfer 25 ml of filtrate to a 250-ml Erlenmeyer flask.

Add 50 ml of $ZnCl_2$ solution and swirl. If the solution remains green or turns yellow, the scrubber contains no free NaOH.

If the solution turns blue, titrate with standard 0.5 *N* HCl to the first definite yellow that persists for 2 s.**

$$\%\ \text{free NaOH} = \frac{\text{titration} \times N \times 40}{\text{sp. gr.} \times \text{ml of sample in aliquot} \times 10}$$

XI.4 AVAILABLE ACID IN CHEMICAL CLEANING SOLUTIONS

As chemical cleaning with an acid proceeds, hydrogen ions are consumed by such reactions as Eqs. (XI-17) and (XI-18), making it necessary to add fresh inhibited acid from time to time to maintain the desired working concentration.

$$Fe_2O_3 + 6H^+ = 2Fe^{+3} + 3H_2O \qquad \text{(XI-17)}$$

$$FeS + 2H^+ = Fe^{++} + H_2S \qquad \text{(XI-18)}$$

* As the specific gravity of a 25 percent solution of NaOH is about 1.27, a significant error is introduced if it is assumed to be 1.00 in the calculation.

** Mercaptans are easily recognized by their characteristic odor; if this is detected, add 10 ml of $CdCl_2$ solution, and mix, before titrating.

For this reason, it is customary to determine the concentration of acid in a recirculating cleaning solution every 30 min so that the strength of the solution can be adjusted. The results also indicate when the cleaning is complete, or at any rate, when no further reaction with acid is taking place, i.e., when the concentration of acid becomes constant from one determination to the next.

The usual laboratory procedure for determining the concentration of an acid in terms of percent, by weight, is to titrate a measured volume of solution with standard alkali using a suitable indicator. The specific gravity of the solution is also determined to find the weight of the volume taken for titration. By noting that the relationship between percent, by weight, and grams per liter is more-or-less linear, a table can be developed containing terms from which percent, by weight, can be calculated from grams per liter, obviating the necessity for determining specific gravity. To show how this is done, consider two solutions of sulfamic acid, one of them containing 105.0 g of HSO_3NH_2 per liter, or 9.95 percent, by weight, and the other containing 20.1 g/l, corresponding to 2.00 percent, by weight. Assuming the relationship between the two sets of values to be linear, an equation for evaluating percent, by weight, in terms of grams per liter can be derived by analytic geometry. Thus, the slope of the straight line defining the relationship between the two terms is:

$$m = (105.0 - 20.1)/(9.95 - 2.00) = 10.68$$

If y is percent, by weight, and x is grams per liter then:

$$y - 105.0 = 10.68(x - \ \ .95)$$

$$= 10.68x - 106.3 + 105.0$$

and

$$x = (y + 1.30)/10.68$$

Terms for an equation of the general form:

$$x = (y + a)/b \qquad \text{(XI-19)}$$

can be evaluated for each of the common acids; this has been done and the results are shown in Table XI-4, which also lists equivalent weights and appropriate indicators for each acid.

TABLE XI-4

Analytical Data for Cleaning Acids

Acid	*Formula*	*Indicator*	*Equivalent weight*	*a*	*b*
Hydrochloric	HCl	Phenolphthalein	36.5	1.00	10.57
Sulfuric	H_2SO_4	Phenolphthalein	49.0	1.30	10.79
Phosphoric	H_3PO_4	Thymolphthalein	49.0	1.10	10.64
Sulfamic	HSO_3NH_2	Phenolphthalein	97.1	1.30	10.68
Citric	$C_7H_8O_6 \cdot H_2O$	Thymolphthalein	70.0	0.80	10.47
Formic	$HCOOH$	Phenolphthalein	46.0	0.50	10.30
Hydroxyacetic	$HOCH_2COOH$	Phenolphthalein	76.0	*	*

* As this acid is customarily used at a concentration of 2 percent combined with formic acid at 1 percent, the specific gravity of the dilute mixture is not significant. The average equivalent weight of the combination is 66.0.

To illustrate the use of Table XI-4, suppose that a 10.0-ml portion of a solution of phosphoric acid required 38.4 ml of 0.5 *N* NaOH to titrate it to the blue end point of thymolphthalein. With this indicator, the equivalent weight of the acid is 49.0.

$$H_3PO_4\text{, g/l} = (38.4\text{ ml} \times 0.5\text{ eq/l} \times 49.0\text{ g/eq})/10\text{ ml}$$
$$= 94.1$$
$$H_3PO_4\text{, percent by weight} = (94.1 + 1.10)/10.64$$
$$= 8.9$$

Note that if specific gravity is neglected, the apparent value of percent, by weight, is 9.4. With any acid weight/volume percentage is, of course, always higher than weight/weight.

Hydrochloric, sulfamic, formic, and hydroxyacetic acids are titrated as monobasic acids and sulfuric acid is titrated as a dibasic acid to the phenolphthalein end point by standard alkali solution. Phosphoric acid is titrated as a dibasic acid and citric acid as a tribasic acid to thymolphthalein indicator. The three dissociation constants of the latter are so similar in value that in a pH titration a single point of inflection is obtained at about pH

10, for which reason thymolphthalein is the indicator of choice; it changes from colorless to blue in the pH range of 9.4-10.6.

When titrating free acid in chemical cleaning solutions account must be taken of the presence of hydrolyzable metal ions, notably ferric and aluminum, that behave as acids toward the indicators specified. Both ferric and aluminum ions form hexafluoro complex ions with fluoride ion, and can thus be rendered innocuous:

$$Fe^{+3} + 6F^{-} = FeF_6^{-3} \qquad \text{(XI-20)}$$

AVAILABLE ACID IN SOLUTIONS TEST

Reagent: *Potassium Fluoride Solution.* Dissolve 30 g of $KF \cdot 2H_2O$ in 100 ml of water and store in a polyethylene bottle.

Procedure: Transfer by means of a pipet a 10-ml portion of filtered cleaning solution to a 250-ml Erlenmeyer flask containing about 100 ml of water.

Add 20 ml of 30 percent KF solution, swirl, and titrate, using the appropriate indicator, with standard 0.5 *N* NaOH. Record the titration and calculate grams per liter and percent, by weight, of acid.*

$$\text{Acid, g/l} = (\text{titr.}\, N \times \text{EW/ml sample})$$

$$\text{Acid, percent by weight} = (\text{g/l} + a)/b$$

XI.5 FREE ALKALI IN CHEMICAL CLEANING SOLUTIONS

Alkaline deoiling solutions are prepared by dissolving a strong base such as NaOH, or salts of weak acids such as Na_2SiO_3, Na_3PO_4, or Na_2CO_3 in water. The hydrolysis of these salts releases hydroxyl ions, which can be determined acidimetrically by titrating to phenolphthalein (P-alkalinity) or thymolphthalein indicator:

$$SiO_3^{--} + H_2O = HSiO_3^{-} + OH^{-} \qquad \text{(XI-21)}$$

$$PO_4^{-3} + H_2O = HPO_4^{--} + OH^{-} \qquad \text{(XI-22)}$$

* Select the indicator and values of *a*, *b*, and equivalent weight listed in Table XI-4 for the acid being determined.

$$CO_3^{--} + H_2O = HCO_3^{-} + OH^{-} \qquad \text{(XI-23)}$$

By continuing the titration to methyl orange (M-alkalinity) or to the mixed indicator bromocresol green–methyl red, two values are obtained from which, in some combinations, the individual concentrations of the above chemicals can be estimated. As applied to solutions containing both sodium hydroxide and sodium carbonate, the alkalinity relationships are as follows:

P-alkalinity (pH 8.3)

$$OH^{-} + H^{+} = H_2O$$

$$CO_3^{--} + H^{+} = HCO_3^{-}$$

M-alkalinity (pH 3.9)

$$OH^{-} + H^{+} = H_2O$$

$$HCO_3^{-} + H^{+} = H_2CO_3$$

$$CO_3^{--} + 2H^{+} = H_2CO_3$$

In Table XI-5 are summarized the relationships between P and M values and the concentrations of hydroxide, carbonate, and bicarbonate.

TABLE XI-5

Alkalinity Relationships

	ppm $CaCO_3$		
Condition	(OH^{-})	(CO_3^{--})	(HCO_3^{-})
$P = 0$	0	0	M
$P < M/2$	0	$2P$	$M - 2P$
$P = M/2$	0	$2P$	0
$P > M/2$	$2P - M$	$2(M - P)$	0
$P = M$	M	0	0

Solutions of sodium silicate can be titrated fairly exactly to the mixed indicator. One mole of silicate consumes two equivalents of acid. Solutions of trisodium phosphate can be analyzed approximately by titrating to thymolphthalein and then to the mixed indicator. The difference between these titrations is used to calculate the concentration of trisodium phosphate. Methods for analyzing the mixtures $NaOH$-Na_2CO_3 and $NaOH$-Na_3PO_4 are prescribed below. These formulations are often used as boil-out solutions for new or retubed boilers, as is a mixture containing 0.4 percent $Na_2HPO_4 \cdot 2H_2O$ and 0.8 percent $Na_3PO_4 \cdot 12H_2O$. The latter combination cannot be conveniently analyzed acidimetrically, but it is useful to know that when the combination is correct the P-alkalinity of the solution will be about 1200 ppm $CaCO_3$ and the M-alkalinity will be about 3200 ppm $CaCO_3$.

XI.5.a Alkaline Deoiling and Boil-Out Solutions

TEST

Reagent: *Mixed Indicator.* Prepare as in Section XI.1.c.

Procedure: Filter the alkaline solution and transfer a measured portion of it, usually 10 ml is appropriate, to a 250-ml Erlenmeyer flask containing about 100 ml of water.

Mixture: Sodium Hydroxide and Trisodium Phosphate

Titrate to thymolphthalein indicator with standard 0.5 N HCl, recording the titration as T ml.

Add mixed indicator and continue the titration to the red end point, recording the total titration as M ml.

$$\text{Percent NaOH} = [(2T - M) \times N \times 40] / [\text{ml sample} \times 10]$$

$$\text{Percent } Na_3PO_4 \cdot 12H_2O = [2(M - T) \times N \times 190] / [\text{ml sample} \times 10]$$

Mixture: Sodium Hydroxide and Sodium Carbonate

Titrate to phenolphthalein indicator with standard 0.5 N HCl, and record the titration as P ml.

Add mixed indicator and continue the titration to the red end point, recording the total titration as M ml.

$$\text{Percent NaOH} = [(2P - M) \times N \times 40] / [\text{ml sample} \times 10]$$

$$\text{Percent } Na_2CO_3 = [2(M - P) \times N \times 53] / [\text{ml sample} \times 10]$$

Sodium Silicate Solutions

Titrate to the mixed indicator with standard 0.5 N HCl, recording the titration as M ml.

$$\text{Percent } Na_2SiO_3 \cdot 9H_2O = [M \times N \times 142] / [\text{ml sample} \times 10]$$

Trisodium Phosphate Solutions

Titrate to thymolphthalein indicator with standard 0.5 N HCl, recording the titration as T ml.

Add the mixed indicator and continue the titration to the red end point, recording the total titration as M ml.

$$\text{Percent } Na_3PO_4 \cdot 12H_2O = [2(M - T) \times N \times 190] / [\text{ml sample} \times 10]$$

Sodium Carbonate Solutions

Titrate to phenolphthalein indicator with standard 0.5 N HCl, recording the titration as P ml.

Add the mixed indicator and continue the titration to the red end point, recording the total titration as M ml.

$$\text{Percent } Na_2CO_3 = [2P \times N \times 53] / [\text{ml sample} \times 10]$$

$$\text{Percent } NaHCO_3 = [(M - 2P) \times 84] / [\text{ml sample} \times 10]$$

XI.6 ANALYSIS OF TURBINE WASHINGS

The types of deposits found in steam turbines and procedures for removing them are discussed on p. 153 and 155, respectively. The most commonly encountered fouling, caused by contaminated steam, is a mixture of boiler salines including sodium chloride, sodium sulfate, and various forms of silica and silicates, especially sodium disilicate. The latter dissolves completely in water and is so strongly hydrolyzed that it titrates quantitatively as NaOH.

$$Na_6Si_2O_7 + 3H_2O = 6Na^+ + 6OH^- + 2SiO_2 \qquad \text{(XI-24)}$$

The progress of a turbine washing can often be followed by periodically determining the concentration of chloride or sulfate in the wash water, but if a silicate is present that requires an alkaline wash to remove, the concentration of silica in the wash solution is more indicative of the efficacy of the washing. Procedures are given below for determining NaCl, Na_2SO_4, and SiO_2 in turbine washings.

The Mohr method for chloride is based on the differential precipitation by silver nitrate of chloride ion and chromate ion, the latter indicating the end point by precipitating as the brick-red Ag_2CrO_4 after all of the chloride has precipitated as the less-soluble AgCl. As the sensitivity of the indicator is markedly decreased in acidic solutions, the pH should be greater than 7, but less than 10.5. For convenience, a factor normality of silver nitrate is used so that if a 100-ml sample is used it is only necessary to multiply the buret reading by 10 to obtain ppm NaCl directly. Because these solutions are dilute, a fairly large indicator blank is obtained, making the determination of small concentrations of sodium chloride imprecise; turbine washings contain 5–250 ppm NaCl.

Sulfate salts, in general, dissolve more slowly than chlorides or sodium hydroxide, so sulfate concentration is frequently measured to monitor the washing of turbine blades to free them from salt deposits. For this type of control the rapid turbidimetric determination specified here is most convenient. As only two reagents are used, and the suspension of barium sulfate formed is stable for almost 10 min, the procedure is simpler than most turbidimetric methods. The usual range is 15–250 ppm SO_4^{--}.

Silica, normally in the range of 10–150 ppm SiO_2, is determined colorimetrically by first forming the yellow complex silico-12-molybdate, $SiMo_{12}O_{40}^{-4}$, by reacting silicate with ammonium molybdate, then reducing the complex to molybdenum blue, $Mo_3O_8 \cdot H_2O$, with a mixed sulfite reducing agent.

XI.6.a Sodium Chloride

TEST

Reagents: *Standard Silver Nitrate Solution* (0.0171 *N*; 1 ml = 1.00 mg NaCl.) Catalog No. HL 2830-4, Hartman-Leddon Company, 60th and Woodland Avenue, Philadelphia, Pennsylvania 19143.
Potassium Chromate Indicator. Dissolve 5 g of K_2CrO_4 in 100 ml of water.

Procedure: Transfer 100 ml of turbine wash water to a Coors No. 3A porcelain casserole.

Add 1–2 drops of phenolphthalein indicator, and if the solution is pink, add 0.02 *N* H_2SO_4 slowly until the solution is colorless.

If the solution is initially colorless, add about 0.2 g of solid $NaHCO_3$ to adjust the pH to about 8.

Add 1.0 ml of K_2CrO_4 indicator solution and titrate with 0.0171 *N* $AgNO_3$ solution to a faint orange end point.

Make a blank determination, substituting distilled water for the sample. This will amount to about 0.5 ml.

$$\text{ppm NaCl} = (\text{titr.} - \text{blank titr.}) \times 10$$

XI.6.b Sodium Sulfate

TEST

Reagents: *Stabilizing Solution.* Dissolve 135 g of $CaCl_2 \cdot 2H_2O$ in about 200 ml of water, then add 50 ml of HCl. Add 500 ml of diethylene glycol (2,2′-oxydiethanol, Eastman No. 2041), and dilute to 1000 ml. Filter if not clear.

Barium Chloride Crystals, 20–30 Mesh. $BaCl_2 \cdot 2H_2O$, J. T. Baker No. 0974.

Standard Sulfate Solution. Dissolve 1.479 g of anhydrous Na_2SO_4 in water, and dilute to 1000 ml (1 ml = 1.00 mg SO_4^{--}).

Standardization. Transfer 0.5, 1.0, 2.0, 3.0, and 4.0-ml portions of standard sulfate solution to a series of 50-ml volumetric flasks, and include a blank.

Add sufficient water to each flask to make 30 ml, then add 10 ml of stabilizing solution, dilute to volume, and mix.

To each flask add about 0.3 g of $BaCl_2$ crystals, stopper, and mix by inverting the flask several times until the crystals are dissolved. Avoid shaking air bubbles into the solution.

Between 5 and 10 min after adding the $BaCl_2$ measure the transmittances of the suspensions at 420 nm in a spectrophotometer.

Plot the observed transmittance against the corresponding milligrams of sulfate on semilogarithmic graph paper.

Procedure: Transfer a filtered sample of turbine washings containing 0.5–4.5 mg of SO_4^{--} to a 50-ml volumetric flask, and dilute to 40 ml with water.

Add 10 ml of stabilizing solution, dilute to volume, and mix.

To the flask add about 0.3 g of $BaCl_2$ crystals, stopper, and mix by inverting the flask several times until the crystals are dissolved.

Measure the transmittance of the solution at 420 nm 5-10 min after adding the barium chloride, then read the corresponding mg of SO_4^{--} from the previously prepared calibration curve.

$$\text{ppm } SO_4^{--} = [\text{mg } SO_4^{--} \times 10^6] / [\text{ml sample} \times 10^3]$$

$$\text{ppm } Na_2SO_4 = \text{ppm } SO_4^{--} \times 1.48$$

XI.6.c Silica

TEST

Reagents: *Ammonium Molybdate Solution.* Dissolve 75 g of $(NH_4)_6$-$Mo_7O_{24} \cdot 4H_2O$ in 800 ml of water, add 60 ml of H_2SO_4, and dilute to 1000 ml.
Sodium Citrate Solution. Dissolve 430 g of $Na_3C_6H_5O_7 \cdot 2H_2O$ in water, and dilute to 1000 ml.
Sulfite Reducing Reagent. Dissolve 80 g of $Na_2S_2O_5$ (sodium metabisulfite) in 800 ml of water. Dissolve 7 g of Na_2SO_3 and 1.5 g of 1-amino-2-naphthol-4-sulfonic acid (Eastman No. 360) in 100 ml of water; stir until dissolved, then combine with the first solution and dilute to 1000 ml with water. (Discard this reagent if it gives a green instead of a blue color in the procedure below.)
Standard Silica Solution. Dissolve 4.8 g of $Na_2SiO_3 \cdot 9H_2O$ in water and dilute to 1000 ml. Standardize gravimetrically, and adjust so that 1 ml = 1.00 mg SiO_2.

Standardization: Transfer 25.0 ml of the standard silica solution to a 100-ml volumetric flask, and dilute to the mark with water (1 ml = 0.25 mg SiO_2).

Transfer 1.0, 2.0, 3.0, 4.0, and 5.0 ml of the dilute standard to a series of 100-ml volumetric flasks, and include a blank.

Dilute the contents of each flask to about 50 ml, then add to each, 5 ml of ammonium molybdate solution, and let stand 5 min.

To each flask add 20 ml of sodium citrate solution while swirling, followed immediately by 5 ml of reducing reagent.

Dilute to the mark with water, mix, let stand 7-10 min, then determine the transmittances of the solutions at 650 nm.

Plot the transmittances observed against the corresponding number of milligrams of silica on semilogarithmic graph paper.

Procedure: Measure a suitable sample of turbine washings and transfer it to a 100-ml volumetric flask, diluting to 50 ml, if necessary. (For wash water containing 0–30 ppm SiO_2 use 50 ml; 10–60 ppm use 25 ml; 50–300 ppm use 5 ml of sample.)

Add 5 ml of molybdate solution and let stand 5 min.

While swirling, add 20 ml of sodium citrate solution, followed by 5 ml of reducing reagent.

Dilute to the mark with water, mix, let stand 7–10 min, then determine the transmittance of the solution at 650 nm.

From the prepared calibration curve read the corresponding mg of SiO_2, correct for a reagent blank, if necessary, and calculate the ppm SiO_2 in the turbine washings.

$$\text{ppm } SiO_2 = (\text{mg } SiO_2 \times 1000)/\text{ml sample}$$

XI.7 RESIDUAL CHELANT IN CHEMICAL CLEANING SOLUTIONS

This procedure is intended to be used for following the course of alkaline cleaning of boilers with ammoniacal EDTA, as described on pp. 81–83. During the first stage of this cleaning procedure, residual chelant is monitored to make sure that an excess of EDTA is present. The cleaning solution in the boiler is then allowed to cool, after which air is blown into the boiler, and the dissolution of copper commences. During the latter stage, residual chelant is also monitored.

Residual chelant is estimated by titrating a portion of the cleaning solution with a standard solution of strontium chloride in the presence of oxalate at a pH of 9.0–9.5. Excess chelant sequesters strontium ion until an excess is present, when the solution becomes turbid with a precipitate of strontium oxalate.

XI.7.a Estimation of Residual Chelant

TEST

Reagents: *Ammonium Oxalate–Saturated Solution.* Dissolve 60 g of $(NH_4)_2C_2O_4 \cdot H_2O$ in 1000 ml of water.

Strontium Chloride Standard Solution. (0.50 *M*) Dissolve 133 g of $SrCl_2 \cdot 6H_2O$ in 800 ml of water, and dilute to volume in a 1000-ml volumetric flask.

Procedure: Transfer a 25-ml sample of cleaning solution to a 250-ml Erlenmeyer flask containing 100 ml of water, and add 5 ml of saturated ammonium oxalate solution.

Titrate the solution with 0.50 *M* $SrCl_2$ solution to the first permanent turbidity.

Check the pH with test paper and if the pH is less than 9.0 add NH_4OH until it is in the range of 9.0–9.5.

If the precipitate dissolves, continue titrating with $SrCl_2$ until a permanent turbidity is formed, then record the titration.

$$\text{Percent } Na_4EDTA = [\text{titr.} \times N \times 380] / [\text{ml sample} \times 10]$$

XI.8 CHROMIC ACID IN CHEMICAL CLEANING SOLUTIONS

Chromic acid is seldom used in industrial chemical cleaning except for oxidizing pyrite or organic binders that hold small carbon particles together in cokes. Its use must be restricted to carbon steel or stainless steel vessels. Rigid steel piping is also required, as hoses are rapidly destroyed by the hot (200°F), 10–20 percent chromic acid solutions customarily applied. This powerful oxidant must never be used in circumstances where it is impossible to obtain a sample of fouling for a solubility and reactivity test.

When chromic anhydride, CrO_3, dissolves in water it immediately hydrolyzes and dissociates to dichromic acid:

$$2CrO_3 + H_2O = 2H^+ + Cr_2O_7^{--} \qquad \text{(XI-25)}$$

which, when hot, oxidizes pyrite, FeS_2, to ferric ion and sulfate:

$$5Cr_2O_7^{--} + FeS_2 + 38H^+ = 10Cr^{+3} + 2Fe^{+3} + 4SO_4^{--} + 19H_2O \qquad \text{(XI-26)}$$

The oxidation of carbonaceous material can be represented as:

$$2Cr_2O_7^{--} + 3C + 16H^+ = 4Cr^{+3} + 3CO_2 + 8H_2O \qquad \text{(XI-27)}$$

Note that in both of these reactions considerably more hydrogen ion is required than is provided by dichromic acid alone; this is the reason that the familiar laboratory cleaning solution is prepared with an excess of sulfuric acid. As the reaction represented by Eq. (XI-27) is highly exothermic, chromic acid must never be applied to easily oxidizable organic material nor to a fouling deposit whose reactivity is unknown. Provision must also be made for venting the carbon dioxide formed to avoid dangerous over-pressuring of the chemical tank and the vessel being cleaned.

The concentration of chromic acid in the range of 5-20 percent can be determined by the iodometric method specified in the procedure below. This method depends upon the oxidation of iodide to iodine by dichromate in acidic solution. The iodine thus produced is then titrated with standard sodium thiosulfate solution.

$$14H^+ + 9I^- + Cr_2O_7^{--} = 3I_3^- + 2Cr^{+3} + 7H_2O \qquad \text{(XI-28)}$$

$$2S_2O_3^{--} + I_3^- = S_4O_6^{--} + 3I^- \qquad \text{(XI-29)}$$

Chemicals that oxidize iodide in acidic solution will, of course, interfere, causing high results. The most likely of these in a chemical cleaning solution is ferric ion, but its interference is easily eliminated by adding ammonium bifluoride to form the hexafluoroferric complex, which does not oxidize iodide.

If there is a possibility of the presence of copper in a sample of chromic acid cleaning solution it can be removed by making the solution alkaline to phenolphthalein with dilute sodium hydroxide, warming to coagulate the precipitate, and filtering. Chromate can then be determined by the procedure below, although in the presence of much iron a small amount of chromate is carried down with the precipitate. If copper is left in the solution it oxidizes iodide according to the following reactions:

$$2Cu^{++} + 5I^- = 2CuI + I_3^- \qquad \text{(XI-30)}$$

or, in the presence of a large excess of iodide:

$$2Cu^{++} + 7I^- = 2CuI_2^- + I_3^- \qquad \text{(XI-31)}$$

It should also be noted that the reaction

$$O_2 + 4I^- + 4H^+ = 2I_2 + 2H_2O \qquad \text{(XI-32)}$$

is catalyzed by copper ions, producing a recurring end point. A cloudy solution after completion of the titration is also an indication of the presence of copper.

XI.8.a Available Chromic Acid

TEST

Reagent: *Thyodene Indicator.* Catalog No. T-138, Fisher Scientific Company, Fair Lawn, New Jersey 07410.

Procedure: Transfer a 5.0-ml portion of chromic acid solution to a 100-ml volumetric flask, dilute to the mark with water, and mix well.

Transfer a 10.0-ml aliquot of the prepared dilution to a 500-ml Erlenmeyer flask containing 100 ml of water. Add 2 ml of 10 N H_2SO_4, 1 g of NH_4HF_2, and swirl to dissolve.

Add 2 g of KI, swirl once, then let stand undisturbed for 5 min, preferably in the dark.

Titrate the solution with standard 0.1 N thiosulfate solution until the yellow triiodide color becomes faint, then add about 0.5 g of Thyodene indicator, and titrate to the disappearance of the blue color:

$$\text{g/l } CrO_3 = (\text{titr.} \times N \times 33.3)/\text{ml sample}$$

$$\text{Percent } CrO_3 = \frac{\text{g/l} + 17.4}{12.5} \quad *$$

XI.9 DETERMINATION OF COPPER

Because of its high thermal conductivity, copper finds frequent use in surface condensers, preheaters, stage heaters, coolers, and various other types of equipment used for heat exchange. Alloys of copper such as manganese brass (often called manganese bronze) and phosphor bronze are used where resistance to corrosion by salt water is required as, for example, for pump impellers in cooling systems using sea water. Tube bundles in heat exchangers are often fabricated of admiralty brass, or in exceptionally corrosive processes, of monel. Corrosion of these alloys leads to the presence of copper and copper oxide deposits in cooling systems and boilers that require special chemical cleaning procedures to remove.

The corrosion of copper in air produces the familiar green basic carbonate, malachite:

* See pp. 232–233 for the derivation of this formula.

$$2Cu + O_2 + CO_2 + H_2O = Cu_2(OH)_2CO_3 \quad \text{(XI-33)}$$

Ammonia and hydrogen cyanide are both especially corrosive to copper because of the formation of very stable complex ions:

$$Cu + 4NH_3 + \tfrac{1}{2}O_2 + H_2O = Cu(NH_3)_4^{++} + 2OH^- \quad \text{(XI-34)}$$

$$Cu + 3HCN = H_2Cu(CN)_3 + \tfrac{1}{2}H_2 \quad \text{(XI-35)}$$

Metallic copper dissolves in the very weak acid HCN ($K_a = 4 \times 10^{-10}$) with the evolution of hydrogen because of the extraordinary stability ($K = 5 \times 10^{-28}$) of the cuprous cyanide ion. This is of no significance in ordinary steam-water equipment, but must be considered in the catalytic cracking of petroleum, in which equipment for recovering hydrogen cyanide is necessary. The following procedures, however, are for measuring the completeness of removal of copper plates and deposits in the chemical cleaning of boilers.

A certain amount of copper is always brought into solution by ferric ion during the acidic cleaning of boilers, as described in Chapter VII, pp. 142-150.

$$2Fe^{+3} + Cu = 2Fe^{++} + Cu^{++} \quad \text{(XI-36)}$$

In order to keep cupric ion in solution when cleaning steel vessels with acid, complexing agents are used, e.g., thiourea (see pp. 150-152). The procedure in Section XI.9.a, below, can be used to judge whether this addition is necessary. If a chelant is used to dissolve copper, as outlined in Chapter II, pp. 81-83, the procedure in Section XI.9.b is useful, while if oxidizing agents such as bromate, persulfate, perborate, or nitrite have been added to solubilize metallic copper (pp. 79-80), the procedure in Section XI.9.c is appropriate. All of these procedures prescribe the iodometric method in which cupric ion oxidizes potassium iodide to iodine, which is then titrated with standard thiosulfate to the starch end point.

$$Cu^{++} + 7I^- = 2CuI_2^- + I_3^- \quad \text{(XI-37)}$$

Ammonium bifluoride is added to each procedure to prevent interference by ferric ion: in Section XI.9.b, EDTA or thiourea are destroyed by boiling a portion of cleaning solution with ammonium persulfate; in Section XI.9.c, sodium sulfite is added to reduce any oxidizing agents that would otherwise interfere.

XI.9.a Copper in Acid Solutions

TEST

Reagent: *Thyodene Indicator.* Catalog No. T-138, Fisher Scientific Company, Fair Lawn, New Jersey 07410.

Procedure: Transfer a 10.0-ml sample of cleaning solution to a 250-ml Erlenmeyer flask containing 100 ml of water.

Add NH_4OH dropwise until a brown precipitate of $Fe(OH)_3$ forms, then add 10 ml of acetic acid.

Add 1 g of NH_4HF_2, stir until dissolved, then add 3 g of KI.

Titrate with standard 0.1 N $Na_2S_2O_3$ until the yellow color of triiodide becomes faint, then add 0.5 g of Thyodene indicator, and titrate until the blue color disappears.* Record the titration.

$$\text{Percent copper} = (\text{ml } S_2O_3^{--} \times N \times 63.5)/(\text{ml sample} \times 10)$$

XI.9.b Copper in Chelant Solutions

TEST

Reagent: *Thyodene Indicator.* As in Section XI.9.a.

Procedure: Transfer a 10.0-ml sample of cleaning solution to a 250-ml Erlenmeyer flask containing 100 ml of water, add 2 g of $(NH_4)_2S_2O_8$, heat to boiling, and boil 10 min, then cool to room temperature.

Add NH_4OH dropwise until a brown precipitate of $Fe(OH)_3$ forms, then add 10 ml of acetic acid.

Add 1 g of NH_4HF_2, stir until dissolved, then add 3 g of KI.

Titrate with standard 0.1 N $Na_2S_2O_3$ until the yellow color of triiodide becomes faint, then add 0.5 g of Thyodene indicator, and titrate to the disappearance of the blue color. Record the titration.

$$\text{Percent copper} = (\text{ml } S_2O_3^{--} \times N \times 63.5)/(\text{ml sample} \times 10)$$

XI.9.c Copper in Oxidizing Solutions

TEST

Reagents: *Thyodene Indicator.* As in Section XI.9.a.

Sodium Sulfite Solution. Dissolve 10 g of Na_2SO_3 in 100 ml of water.

* Some corrosion inhibitors impart a yellow cast to the solution.

Procedure: Transfer a 10.0-ml sample of cleaning solution to a 250-ml Erlenmeyer flask, and place the flask in a fume hood.

Add 5 ml of 10 percent Na_2SO_3 and 10 ml of 1:1 HCl, and boil gently until the odor of SO_2 is undetectable.

Wash down the walls of the flask with water, diluting the contents to about 50 ml. Cool to room temperature.

Add NH_4OH dropwise until a brown precipitate of $Fe(OH)_3$ forms, then add 10 ml of acetic acid.

Add 1 g of NH_4HF_2, stir until dissolved, then add 3 g of KI.

Titrate with standard 0.1 N $Na_2S_2O_3$ until the yellow color of triiodide becomes faint, then add 0.5 g of Thyodene indicator, and titrate to the disappearance of the blue color. Record the titration.

$$\text{Percent copper} = (\text{ml } S_2O_3^{--} \times N \times 63.5)/(\text{ml sample} \times 10)$$

XI.10 DETERMINATION OF IRON

Deposits of hematite and magnetite are very common and, indeed, often predominate in boiler and other water-formed deposits; the solubilizing effect of any acid on these oxides is represented by the following equations:

$$Fe_2O_3 + 6H^+ = 2Fe^{+3} + Fe^{++} + 3H_2O \qquad \text{(XI-38)}$$

$$Fe_3O_4 + 8H^+ = 2Fe^{+3} + Fe^{++} + 4H_2O \qquad \text{(XI-39)}$$

By determining the concentration of total iron in a cleaning solution every 30 min, or so, during the course of the operation, the chemical cleaning contractor can judge when to terminate the acidification step, viz., when the concentration of total iron in the solution reaches a reasonably constant value. Ferric ion, from whatever source, rapidly oxidizes iron to ferrous ion because of the very large change in free energy that occurs in the reaction.

$$2Fe^{+3} + Fe = 3Fe^{++} \qquad \text{(XI-40)}$$

The enormous equilibrium constant of this reaction ($K = 1 \times 10^{41}$) indicates that the concentration of ferric ion that can exist in equilibrium with a steel surface is vanishingly small and that, therefore, there is no purpose in trying to measure its concentration. As many people think this is significant, however, a procedure is supplied in Section XI.10.b that could determine the concentration of ferric ion, were any present. Note that copper interferes in all three procedures given here for determining iron, and account must be taken of its concentration as measured by any appropriate procedure in Section XI.9.

The net reactions of hematite and magnetite with hydrogen ions and steel are:

$$Fe_2O_3 + Fe + 6H^+ = 3Fe^{++} + 3H_2O \qquad \text{(XI-41)}$$

$$Fe_3O_4 + Fe + 8H^+ = 4Fe^{++} + 4H_2O \qquad \text{(XI-42)}$$

A consideration of molecular weights in these reactions shows that 160 lb of hematite oxidizes 56 lb of boiler steel (one-third of the total) and 232 lb of magnetite oxidizes 56 lb of boiler steel (one-fourth of the total). It is of interest to calculate the effect of this on corrosion rate. Assume, for illustration, that a steel boiler containing 20,000 ft^2 of surface is cleaned with 15,000 gal of 5 percent hydrochloric acid, which at the conclusion of the operation is found to contain 2.3 percent of total iron. The density of a 5 percent solution of HCl is 8.53 lb/gal (see Table II-2, p. 46) and the density of steel is 490 lb/ft^3 (see Table I-8, p. 25). Therefore,

$$\text{total pounds Fe} = 15{,}000\ \text{(gal)} \times 8.53\ \text{(lb/gal)} \times 0.023 = 2943$$

Assuming uniform attack on the steel boiler, and that the deposit in it is all hematite, the depth of corrosion is:

$$\frac{0.33 \times 2943\ \text{(lb)} \times 12\ \text{(in./ft)} \times 1000\ \text{(mil/in.)}}{20{,}000\ (\text{ft}^2) \times 490\ (\text{lb/ft}^3)} = 1.19\ \text{mil}$$

If this corrosion occurred during 6 h of acid treatment the corrosion rate would be:

$$1.19 \times 24 \times 365/6 = 1737\ \text{mpy}$$

The rate in hot, inhibited acid (5 percent HCl @ 150°F) in the absence of ferric ion is 250–350 mpy.

XI.10.a Total Iron in Acidic Solutions

TEST

Reagents: *Permanganate Solution.* Dissolve 3.3 g of $KMnO_4$ in 1000 ml of water.

Sodium Azide Solution. Dissolve 5 g of NaN_3 in 100 ml of water.
Thyodene Indicator. Catalog No. T-138, Fisher Scientific Company, Fair Lawn, New Jersey 07410.

Procedure: Transfer a 10.0-ml sample* of cleaning solution to a 250-ml Erlenmeyer flask, dilute to about 40 ml with water, then add $KMnO_4$ solution dropwise until a permanent pink color is obtained.

Add 1 drop of NaN_3 solution to discharge the color, add NH_4OH dropwise until a brown precipitate of $Fe(OH)_3$ forms, then add 10 ml of acetic acid.

Add 3 g of KI, and titrate with standard 0.1 N $Na_2S_2O_3$ until the yellow color of the triiodide becomes faint, then add 0.5 g of Thyodene indicator, and titrate to the disappearance of the blue color. Record the titration.

Percent total iron =

$$[(\text{ml total titr.} - \text{ml copper titr.}) \times N \times 56] / (\text{ml sample} \times 10)$$

XI.10.b Ferric Ion in Acidic Solutions

TEST

Procedure: Proceed as in Section XI.10.a, omitting the addition of $KMnO_4$ and NaN_3, and commencing with "...add NH_4OH dropwise, etc."

Percent ferric ion =

$$[(\text{ml total titr.} - \text{ml copper titr.}) \times N \times 56] / (\text{ml sample} \times 10)$$

XI.10.c Total Iron in Chelant Solutions

TEST

Reagent: *Thyodene Indicator.* As in Section XI.10.a.

Procedure: Transfer a 10.0-ml sample of cleaning solution to a 250-ml Erlenmeyer flask containing 100 ml of water, add 2 g of $(NH_4)_2S_2O_8$,

* In all procedures in this section both copper and iron oxidize iodide to iodine so the total titration here must be corrected by subtracting that obtained in any of the procedures for copper. It is convenient to use the same sample size for both.

heat to boiling and boil 10 min, then cool to room temperature.

Add NH_4OH dropwise until a brown precipitate of $Fe(OH)_3$ forms, then add 10 ml of acetic acid.

Add 3 g of KI and titrate with 0.1 N $Na_2S_2O_3$ until the yellow color of triiodide is faint, then add 0.5 g of Thyodene indicator and titrate to the disappearance of the blue color. Record the titration.

Percent total iron =

$$[(\text{ml total titr.} - \text{ml copper titr.}) \times N \times 56]/(\text{ml sample} \times 10)$$

XI.11 DETERMINATION OF PHOSPHATE

Phosphate salts are included in a number of boil-out and passivating solutions used in chemical cleaning. In an extended boil-out of a new boiler it is necessary to add phosphate from time to time to keep its concentration at the desired level, and for this reason a method for determining the concentration of phosphate in boil-out solutions is included here. The following procedure makes use of the molybdenum blue color produced by the selective reduction of molybdiphosphoric acid, $[H_3P(Mo_3O_{10})_4]$ resulting from the condensation of ortho-phosphate and molybdate ions in weakly acidic solution.

Greenberg, et al.[5] have called attention to the interference by nitrite in this method. As nitrite is a common ingredient in passivating solutions and is a decomposition product of sodium nitrate, used to inhibit stress corrosion cracking by alkaline boil-out solutions, their recommendation of adding sulfamic acid to the molybdate reagent has been followed. The reaction between nitrite (or nitrous acid) and sulfamic acid is as follows:

$$NO_2^- + HSO_3NH_2 = N_2 + H_2O + SO_4^{--} + H^+ \qquad \text{(XI-43)}$$

XI.11.a Ortho-Phosphate in Cleaning Solutions

TEST

Reagents: *Molybdate Solution.* Dissolve 37.5 g of $(NH_4)_6Mo_7O_{24} \cdot 4H_2O$ in about 600 ml of water. Slowly add 300 ml of H_2SO_4, and cool. To the solution add 10 g HSO_3NH_2, stir to dissolve, and dilute to 1000 ml.

Sulfite Reducing Reagent. Dissolve 80 g of $Na_2S_2O_5$ (sodium meta bisulfite) in 800 ml of water. Dissolve 7 g of Na_2SO_3 and 1.5 g of 1-amino-2-

naphthol-4-sulfonic acid (Eastman No. 360) in 100 ml of water; stir until dissolved, then combine with the first solution and dilute to 1000 ml with water. (Discard this reagent if it gives a green instead of a blue color in the procedure below.)

Standard Phosphate Solution. Dissolve 1.433 g of KH_2PO_4 in water and dilute to 1000 ml. (1 ml = 1.00 mg PO_4^{-3}.)

Standardization: Dilute 25.0 ml of phosphate standard solution to 100 ml to produce a solution containing 0.25 mg PO_4^{-3}/ml. Transfer 2.0, 4.0, 6.0, 8.0, and 10.0 ml of the dilute standard, corresponding to 0.5, 1.0, 1.5, 2.0, and 2.5 mg of PO_4^{-3}, to a series of 100-ml volumetric flasks and include a blank.

Bring the total volume of each flask to about 80 ml, then while swirling, add 5 ml of molybdate reagent, followed by 5 ml of sulfite reducing reagent. Dilute to volume and mix thoroughly.

Measure the transmittances of the solutions at 650 nm between 7 and 10 min after adding the second reagent. Plot the results against the corresponding mg of PO_4^{-3} on semilogarithmic graph paper.

Procedure: Filter the sample solution if it is not clear, transfer a 10.0-ml portion to a 100-ml volumetric flask, dilute to volume, and mix.

Transfer a 5.0-ml aliquot* of the prepared solution to a 100-ml volumetric flask, and dilute to about 80 ml.

Add, while swirling, 5 ml of molybdate reagent, followed by 5 ml of sulfite reducing reagent. Dilute to the mark and mix thoroughly.

Measure the transmittance of the solution at 650 nm, 7–10 min after adding the second reagent, and determine the mg of PO_4^{-3} from the prepared calibration curve.

$$\text{ppm } PO_4^{-3} = (\text{mg } PO_4^{-3} \times 10^8)/(\text{ml sample} \times \text{ml aliquot} \times 10^3)$$

XI.12 EVALUATION OF INDUSTRIAL CLEANERS

External cleaning of concrete, tanks, and industrial equipment that tends to become soiled with oil or grease can be conveniently done with commercial cleaners applied, together with steam and water, through a special mixing jet connected to a high-pressure hose. An enormous number of these cleaners is available, but the most effective ones for removing oil

* This aliquot, containing 0.5 ml of the original solution, covers the range of 1000–5000 ppm PO_4^{-3}, which includes most boil-out and passivating solutions.

and grease are alkaline detergents containing combinations of NaOH, Na_3PO_4, Na_2SiO_3, wetting agents, and emulsifiers. In order to screen these cleaners, they are given a brief evaluation in the laboratory, followed by field testing at various dilutions. Because of hazardous waste regulations the pH of a cleaner, diluted, and undiluted, should not exceed 12.5; the wetting time for aqueous cleaners should not exceed 30 s. After making the qualifying evaluations, field tests are conducted at various concentrations ranging from 1:10 to as high as 1:40, after which the cost of the highest effective dilution is calculated.

So many different pH meters are manufactured, and their use is so common, that no attempt to write a detailed procedure for measuring pH will be made. The meter is standardized, however, by immersing the electrodes in a standard buffer solution, and then adjusting the meter to indicate the defined value, pH_s. The electrodes are rinsed thoroughly, then immersed in the unknown solution, the pH of which is indicated by the meter. The reader is referred to an excellent work by Wescott[6] dealing with the practical aspects of making accurate pH measurements.

XI.12.a Wetting Time of Detergents

The following procedure can be used to compare the rate of wetting of various alkaline detergents. To do this a canvas disk is placed on the surface of a 0.5 percent solution of industrial cleaner, and the time in seconds required for the disk to sink to the bottom of the container is recorded as the wetting time.

RATE OF WETTING TEST

Equipment: *Canvas Disks.* 1 in. in diameter, cut from 10-oz cotton duck.

Procedure: Mix the sample of detergent thoroughly, weigh a 2.5-g portion, and transfer it to a 500-ml glass-stoppered mixing cylinder. Dilute to 500 ml, and mix thoroughly.

Pour a sufficient amount of the prepared solution into a 250-ml beaker to obtain a depth of 3 in., and adjust the temperature to 77 ± 2°F.

Remove any bubbles from the surface of the solution, then using forceps, carefully place a canvas disk upon the surface of the solution near the center.

Record the time in seconds for the disk to be wetted and sink to the bottom of the beaker.

Remove the disk, and repeat for a total of 10 tests, using a new disk for each.

$$\text{wetting time} = \text{average time in seconds}$$

References

(1) McCoy, J. W. 1969. *Chemical analysis of industrial water.* New York: Chemical Publishing Co.

(2) Am. Soc. Testing Materials. 1981. Identification of crystalline compounds in water-formed deposits by x-ray diffraction. *1981 Book of ASTM Standards.* Part 31, Method D934.

(3) *Ibid.* 1981. Analysis of water-formed deposits by x-ray fluorescence. Part 31, Method D2332.

(4) *Ibid.* 1981. Comparison of waterborne petroleum oils by gas chromatography. *1981 Book of ASTM Standards.* Part 31, D3328.

(5) Greenberg, A. E., Weinberger, L. W., and Sawyer, C. N. 1950. Control of nitrite interference in colorimetric determination of phosphorus. *Anal. Chem.* 22: 499.

(6) Wescott, C. C. 1978. *pH measurements.* New York: Academic Press.

Appendix I

Glossary of Terms

Acidic emulsion. An emulsion of a mineral acid with a solvent that is used to remove fouling deposits composed of both organic and inorganic components.

Acidic foam. A foam prepared by blowing nitrogen into an acidic solution containing a foaming agent; especially useful for cleaning large vessels such as surface condensers and boilers.

Alkaline emulsion. An emulsion of an alkaline solution with a solvent that is used to emulsify and disperse waxy, greasy, or polymeric organic fouling.

Alpha-brasses. Alloys (solid solutions) of up to 40 percent zinc in copper.

Anode. The positive electrode of an electrochemical cell where electrons are donated and oxidation occurs.

Atmospheric preheat furnace. A furnace that supplies heat to crude petroleum feed to a distillation column operated at, or near, atmospheric pressure.

Austenite. The face-centered cubic crystalline phase of γ-iron in iron-based alloys such as austenitic stainless steels.

Baffle. A plate or partition installed in various industrial equipment to direct or divert the flow of fluids.

Black liquor. Alkaline sulfide–sulfate solutions for digesting wood in pulp or paper plants.

Blend-filling. The preparation of a chemical cleaning solution by simultaneously metering water and chemical in appropriate proportions into a vessel to be cleaned.

Blind. A solid plate bolted between two pipe flanges to prevent flow through the pipe.

Block valve. A valve designed to be either open or closed, as distinguished from a regulating valve, for example.

Blowdown. A portion of recirculating water that is continuously or intermittently removed from a boiler to limit the concentration of soluble or insoluble material.

Boilermaker blind. Essentially a flanged plate or cap installed on the end of a pipe or header.

Boilermaker staging. Scaffolding with limited working area erected for the use of boilermakers while removing channel covers from heat exchangers and the like.

Boiler tubes. Pipes that transport water and steam within a boiler.

British thermal unit. The quantity of heat required to raise the temperature of one pound of water one degree Fahrenheit.

Bubble-cap tray. A horizontal plate in a fractionating column pierced with holes fitted with chimneys, and covered by movable caps.

Catalytic reforming. A process for converting heavy paraffinic naphtha to aromatics to improve octane number.

Cathode. The negative electrode of an electrochemical cell where electrons are accepted and reduction occurs.

Caustic embrittlement. Cracking of stressed steel in contact with concentrated alkali.

Caustic gouging. Attack on steel by concentrated sodium hydroxide in which hydrogen is evolved and ferrite and hypoferrite ions are formed. Also called ductile gouging.

Channel. The large open passage, usually fitted with pass partitions, through which cooling water is introduced to the tubes in the end of a heat exchanger.

Check valve. A valve that confines the flow of a fluid to one direction.

Chelant. An organic compound capable of forming a cyclic structure in which a metal ion is bound as a coordination complex.

Chelonate. A coordination complex formed by a chelant and a metallic ion.

Compressor jacket. A chamber through which water is circulated to cool the bearings or other parts of a compressor.

Condensate. Water derived from condensed steam.

Contact heater. A device in which water is heated by the direct injection of steam.

Convection. Transmission of heat by propagation through a gas or fluid.

Cooling tower. An enclosed structure for the evaporative cooling of water by contact with a stream of air.

Corrosion. The oxidation of a metal or alloy to its corresponding cations or oxides.

Corrosion coupons. Weighed strips of metal suspended in a flowing stream of cleaning solvent in an attempt to measure its corrosivity.

Corrosion current. The stream of electrons flowing from anodic to cathodic areas through a metal undergoing corrosion.

Corrosion inhibitor. A chemical capable of raising the hydrogen overvoltage on the surface of a metal, thereby slowing the rate of its corrosion.

Corrosion rate. The average rate of penetration of a metal, usually expressed in mils per year (mpy) or milligrams per square decimeter per day (mdd).

Deposit. Any sort of insoluble material laid down in industrial equipment. See also scale and fouling.

Dezincification. A corrosion process in brasses in which the more reactive component (zinc) is selectively oxidized, leaving the metallic surface enriched in the less reactive component (copper).

Dispersant. A chemical that promotes the suspension of insoluble particles.

Distillation column. A device for separating mixtures of liquids into specific boiling ranges.

Ductile gouging. See caustic gouging.

Economizer. A device for transferring heat from flue gas to feed water to a boiler.

EDTA. Abbreviation for ethylenedinitrilotetraacetic acid.

Epitaxial growth. A mechanism of scaling in which a crystalline substance grows on a second crystalline form already present as a substrate. An example of epitaxy is the growth of $(Zn,Fe)_3(PO_4)_2 \cdot 4H_2O$ on a preexisting scale of $Zn_3(PO_4)_2 \cdot 4H_2O$.

Evaporator. A device used on ships to prepare distilled water from sea water.

Feed preheat exchanger. A heat exchanger in which feed stock to a distillation column is heated by passing a side cut from the column through the exchanger.

Filming amine. A nitrogen base added to acidic chemical solutions to inhibit corrosion of the equipment to be cleaned.

Flare. A device for incinerating waste gases.

Flash point. The temperature at which the vapor above a heated portion of a flammable solvent momentarily ignites when a small pilot flame is passed across the surface.

Floating-head heat exchanger. A heat exchanger in which only one tube sheet is welded to the shell; the other, closed by a flanged floating-head cover, floats freely in the shell to accomodate differential thermal expansion.

Flue gas. The product of combustion of fuel in a furnace consisting primarily of water and carbon dioxide, with small amounts of carbon monoxide and sulfur dioxide.

Fluid catalytic cracking. A process for converting heavy gas oil to gasoline and jet fuel.

Fouling. General term for any accumulation of insoluble material that impedes circulation or heat transfer.

Fouling factor. The reciprocal of the heat transfer coefficient multiplied by 1000.

Half-reaction. A chemical equation representing either the oxidation or reduction half of a complete oxidation–reduction equation.

Head. The height to which a liquid can be raised by a given pressure.

Header. A common line into which other lines flow.

Heat exchanger. A tube-in-shell vessel in which heat is transferred between fluids in the tubes and in the shell.

Heat flux. Rate of transfer of thermal energy through a given surface. See heat transfer coefficient.

Heat transfer coefficient. The rate, in Btu/ft^2-h-°F, at which heat flows through a surface.

Hydroblasting. A method of cleaning in which water at high pressure is directed in a narrow jet at the surface to be cleaned.

Hydrocracking. Catalytic cracking of heavy gas oil in the presence of hydrogen to produce heavy gasoline and jet fuel.

Hydrofluoric acid alkylation. The alkylation of light isoparaffins with light olefins to form higher isoparaffins, using HF as a catalyst. Essentially, this process is the reverse of cracking.

Hydrogen bond. A single-electron bond, largely electrostatic, formed between hydrogen atoms in molecules of water.

Hydrogen overvoltage. The potential required to discharge, as hydrogen gas, the layer of adsorbed hydrogen atoms on the surface of a metal.

Inhibitor. One of various chemicals added to cleaning solutions to impede the attack of hydrogen ions on metal surfaces.

Integral-furnace boiler. A steam generator in which both the furnace and boiler are combined in a single structure.

Magnetite. The film of Fe_3O_4 formed when steel is heated and normally present as a protective film on internal boiler surfaces.

Martensite. A body-centered tetragonal crystalline supersaturated solution of carbon in the iron phase of iron-based alloys.

Mill scale. A mixture of FeO, Fe_2O_3, and Fe_3O_4 formed on steel surfaces at high temperature during fabrication.

Mud drum. Lower drum of a boiler where insoluble sludge collects.

Nipple. A short length of threaded pipe used for connecting a valve, gauge, or fitting to existing piping or equipment.

NTA. Abbreviation for nitrilotriacetic acid.

Oxidation. The addition of oxygen, the removal of hydrogen, or the loss of electrons.

Oxidation potential. The voltage required to remove electrons from a substrate being oxidized.

Passivation. The transformation of a metallic surface to a state in which it is not readily oxidized; usually accomplished by forming a protective film.

Pass partition. Baffles in the channels of heat exchangers that determine the path of water flow through the vessel.

Polarization. The extent of potential change, in volts, caused by a net flow of current to or from an electrode.

Pyrophore. A substance that ignites spontaneously in air.

Radiant heat. Heat supplied by radiation without the necessity of some medium for transmission. (e.g., heat from the sun is radiant heat.)

Reactor. One of many kinds of vessels in which a chemical reaction is brought about.

Reboiler. A vessel containing steam coils in which hydrocarbons or other process chemicals are evaporated.

Reduction. The removal of oxygen, the addition of hydrogen, or the addition of electrons.

Relief system. An arrangement for capturing gases emitted through pressure relief valves and piping them to a flare for incineration.

Saponification. The hydrolysis of a fat to a soap by alkali.

Scale. A hard, adherent, crystalline layer of a slightly soluble salt deposited on a surface through which heat is exchanged.

Sequestrant. General term for any chemical that forms a soluble complex or chelonate with ions that would otherwise form insoluble deposits.

Shell nozzle. The opening through which material passes into or out of the shell of a heat exchanger.

Shotgunning. Water jetting using a hand-held hydroblasting gun.

Sieve tray. A plate in a distillation column perforated with many small holes.

Slag. Molten salts of vitreous consistency that accumulate in furnaces fueled by residual fuel oil.

Slip blind. A plate that slips between two pipe flanges and prevents flow through the pipe.

Sludge. General term for any sort of wet or sticky accumulation in industrial equipment.

Spectacle blind. A two-ended slip blind, one of which is an open steel ring.

Splash packing. A type of fill consisting of staggered rows of slats so arranged that water drops falling through a cooling tower are broken into droplets thus achieving more intimate contact between water and air.

Spool. A short length of pipe, flanged at both ends, used to replace a valve during chemical cleaning.

Square-pitched tube bundle. A collection of tubes so arranged that connecting the centers of any four tubes defines a square.

Stage heaters. Heaters, usually for heating boiler feed water, connected in series so that each is progressively hotter.

Steam-air decoking. A process for removing coke by passing steam and air through hot furnace tubes.

Steam drum. The upper drum in a boiler where feed water and chemicals are injected and blowdown water and saturated steam are withdrawn.

Stripping column. A column in which gases and light ends are separated from other materials by injecting steam.

Superheater. Tubes within a furnace through which saturated steam passes after leaving a boiler to absorb additional heat.

Surface condenser. A water-cooled device for condensing steam.

Surfactant. A chemical that alters surface and interfacial tension of solutions.

Swedge nipple. A reducing nipple, male-threaded on both ends, e.g., from ¾-in. to ½-in.

Tray column. A fractionating column containing a number of trays, one above another, fitted with bubble caps, valves, or pierced with holes.

Triangular-pitched tube bundle. A collection of tubes so arranged that connecting the centers of any three tubes defines an equilateral triangle.

Tube bundle. A collection of tubes that are installed in the shell of a heat exchanger or waste heat boiler.

Tube sheet. A supporting plate drilled to hold an array of tubes as, for instance, in a heat exchanger, feed water heater, or economizer.

Turbining. A mechanical method for cleaning boiler tubes in which rotating cutters are pushed through the tubes, crushing and cutting through scale.

Valve tray. A steel plate with numerous perforations, each fitted with a check valve, used in distillation columns.

Valve trim. Seats, disks, and stems of valves, which are often fabricated from unusual alloys that may be damaged by chemical cleaning.

Vapor phase inhibitor. An inhibitor that volatilizes from hot solutions and protects the vapor space of a vessel from corrosion by a chemical cleaning solution.

Waste heat boiler. A boiler that uses waste heat or low grade fuel to generate steam.

Water wall header. The common line to which water wall tubes are connected.

Water wall tubes. Tubes containing boiler water that shield refractory furnace walls by absorbing heat.

Wetting agent. See surfactant.

Appendix II

Numerical Problems

1. The acidic dissociation constant of acetic acid is $K_a = 1.7 \times 10^{-5}$. What is the pH of a 2 percent solution of the acid in water?

Ans. 2.1

2. The third acidic dissociation constant of phosphoric acid is $K_3 = 4.4 \times 10^{-13}$ and the ion product of water is 1.0×10^{-14}. If 2 g of Na_3PO_4 are dissolved in 200 ml of water, what will be the pH of the solution?

Ans. 12.7

3. Calculate the weight of anhydrous HCl in 5.00 ml of concentrated hydrochloric acid of specific gravity 1.19 and containing 37.23 percent of HCl, by weight.

Ans. 2.215 g HCl

4. Calculate the volume of concentrated H_2SO_4, specific gravity 1.84, containing 98 percent H_2SO_4, by weight, that would contain 40.0 g of anhydrous H_2SO_4.

Ans. 22.2 ml

5. What volume of concentrated H_2SO_4, specific gravity 1.84, containing 98 percent H_2SO_4, must be diluted with water to prepare 100 ml of 20 percent H_2SO_4, specific gravity 1.14?

Ans. 12.7 ml

6. Suppose instead of 100 ml one wished to prepare 100 gal of 20 percent H_2SO_4. How many gallons of 98 percent H_2SO_4 should be diluted to 100 gal with water?

Ans. 12.7 gal

7. What volume of concentrated HCl (37.23 percent, by weight, specific gravity 1.19) and what volume of water are required to prepare 500 gal of 10 percent HCl, by weight (i.e., 50 lb of HCl and 450 lb of water)?

Ans. 118.5 gal HCl
381.5 gal H_2O

8. What volume of 28 percent HCl, specific gravity 1.139, is required to prepare 750 gal of 6 percent HCl, specific gravity 1.028?

Ans. 145 gal HCl

9. A chemical cleaning contractor wishes to prepare 2500 gal of 7 percent HCl, which has a normality of 1.99 equivalents per liter. Upon checking the normality of his prepared solution, however, he finds that it is 2.24 normal. How many gallons of water must be added to obtain the proper concentration?

Ans. 314 gal

10. Using 77.8 percent H_2SO_4, a contractor prepares 675 gal of a solution having a normality of 1.73 equivalents per liter. If he wants a 10 percent solution, how many gallons of 77.8 percent H_2SO_4 must he add?

Ans. 12 gal

11. How many pounds of limestone, $CaCO_3$, are needed to neutralize 355 gal of 4 percent H_2SO_4?

Ans. 124 lb $CaCO_3$

12. How many pounds of chromic anhydride, CrO_3, are needed to prepare 255 gal of 12 percent chromic acid solution?

Ans. 290 lb CrO_3

13. What is the specific gravity of a 7 percent solution of phosphoric acid?

Ans. 1.0365

14. It is proposed to use a proprietary corrosion inhibitor containing 45 percent chromate and 6 percent zinc to passivate a water-cooled heat exchanger after chemical cleaning. (a) If the passivating solution contains 1000 ppm CrO_4^{--} and its pH is 7.2, what is the maximum concentration of zinc that could be in solution assuming that the solubility product of $Zn(OH)_2$ is 1.2×10^{-17}? (b) What percentage of the zinc added will precipitate at pH 7.2?

Ans. (a) 31 ppm Zn
(b) $\approx 77\%$ pptd

15. A contractor proposes to clean a heat exchanger containing an unknown quantity of ferrous sulfide by washing it with 2000 gal of 10 percent HCl. Supposing he uses a single scrubber, containing 600 gal of sodium hydroxide solution, vented to a low-pressure relief system, calculate the following values, using a chemical handbook, as required: (a) The number of pounds of HCl in the 10 percent solution; (b) The number of pounds of NaOH the HCl can neutralize; (c) The number of additional pounds of NaOH needed to ensure a 10 percent excess in the scrubber after all of the acid is neutralized; (d) The total number of pounds of NaOH to add to the scrubber; (e) The number of gallons of 50°Bé NaOH that contains the pounds of NaOH calculated in (d); (f) The total weight of FeS dissolved by 2000 gal of 10 percent HCl; (g) The total volume of H_2S released under standard conditions.

Ans. (a) 1748 lb HCl
(b) 1915 lb NaOH
(c) 555 lb NaOH
(d) 2470 lb NaOH
(e) 388 gal 50°Bé NaOH
(f) 2107 lb FeS
(g) 8600 ft^3 H_2S

Index